At the sea's edge

PHalarope Books

PHalarope Books are designed specifically for the amateur naturalist. These volumes represent excellence in natural history publishing. Each book in the PHalarope series is based on a nature course or program at the college or adult education level or is sponsored by a museum or nature center. Each PHalarope Book reflects the author's teaching ability as well as writing ability.

BOOKS IN THE SERIES

The Curious Naturalist
John Mitchell and The Massachusetts Audubon Society

The Amateur Naturalist's Handbook
Vinson Brown

The Amateur Naturalist's Diary
Vinson Brown

Outdoor Education: A Manual
for Teaching in Nature's Classroom
Michael Link, Director, Northwoods Audubon Center, Minnesota

Nature with Children of all Ages:
Activities and Adventures for Exploring, Learning,
and Enjoying the World Around Us
Edith Sisson, The Massachusetts Audubon Society

The Wildlife Observer's Guidebook
Charles E. Roth, The Massachusetts Audubon Society

The Fossil Collector's Handbook
James Reid Macdonald

Nature in the Northwest: An Introduction to the Natural History
and Ecology of the Northwestern United States
from the Rockies to the Pacific
Susan Schwartz, photographs by Bob and Ira Spring

Nature Drawing: A Tool for Learning
Clare Walker Leslie

Nature Photography: A Guide to Better Outdoor Pictures
Stan Osolinski

The Art of Painting Animals: A Beginning Artist's Guide
to the Portrayal of Domestic Animals, Wildlife, and Birds
Fredric Sweney

A Complete Manual of Amateur Astronomy:
Tools and Techniques for Astronomical Observations
P. Clay Sherrod with Thomas L. Koed

365 Starry Nights: An Introduction
to Astronomy for Every Night of the Year
Chet Raymo

Prentice-Hall, Inc., Englewood Cliffs, New Jersey 07632

A SPECTRUM BOOK

WILLIAM T. FOX
Illustrated by Clare Walker Leslie

AN INTRODUCTION
TO COASTAL OCEANOGRAPHY
FOR THE AMATEUR NATURALIST

At the
sea's edge

Library of Congress Cataloging in Publication Data

Fox, William T.
    At the sea's edge.

    (A PHalarope book)
    "A Spectrum Book."
    Bibliography: p.
    Includes index.
    1. Coasts.   I. Title.
GB451.2.F69   1983   508.314'6   82-23074
ISBN 0-13-049783-5
ISBN 0-13-049775-4 (pbk.)

© 1983 by Prentice-Hall, Inc., Englewood Cliffs, New Jersey 07632

A Spectrum Book

Printed in the United States of America

1 2 3 4 5 6 7 8 9 10

This book is available at a special discount when ordered in bulk quantities.
Contact Prentice-Hall, Inc., General Publishing Division, Special Sales,
Englewood Cliffs, N.J. 07632.

ISBN 0-13-049783-5

ISBN 0-13-049775-4 {PBK.}

Editorial/production supervision by Kimberly Mazur
and Norma G. Ledbetter
Interior design by Kimberly Mazur
Cover design by Hal Siegel
Manufacturing buyer: Cathie Lenard

Prentice-Hall International, Inc., *London*
Prentice-Hall of Australia Pty. Limited, *Sydney*
Prentice-Hall Canada Inc., *Toronto*
Prentice-Hall of India Private Limited, *New Delhi*
Prentice-Hall of Japan, Inc., *Tokyo*
Prentice-Hall of Southeast Asia Pte. Ltd., *Singapore*
Whitehall Books Limited, *Wellington, New Zealand*
Editora Prentice-Hall do Brasil Ltda., *Rio de Janeiro*

To Norma, Steve, Katherine, and Amy, for the many happy hours
we have spent together exploring the coast.

Contents

Preface

Coastal Oceanography—the scientific study of the coast—is a blend of several sciences that are focused on the shore. The shore is treated as a field laboratory where one can observe nature and study forces that shaped the coast. This book is aimed at the amateur naturalist with a good high school background in science, and perhaps a college course in biology or geology. The book is illustrated throughout with drawings, photographs, and field sketches that can be used as a guide to the coast. It can also serve as a text for an advanced high school or introductory college course in coastal oceanography.

Each of the natural sciences makes a contribution toward our understanding of the coast. Geology explains the formation of sea cliffs, beaches, barrier islands, and reefs. The theories of sea floor spreading and plate tectonics are introduced to account for the differences in character between the Atlantic and Pacific coasts. Topics related to meteorology include coastal storms, hurricanes, sea breezes, and coastal fogs. Astronomy and physics are used to explain the behavior of waves, currents, and tides. Ideas from the field of optics are applied to waves breaking along the shore. Marine biology and ecology provide a key to the behavior and distribution of coastal plants and animals that are specially adapted to the intertidal zone. The rocky coast, beach, dune, salt marsh, and coral reef are studied as examples of coastal habitats.

Several simple techniques are introduced for making field observations and measurements along the coast. Clare Walker Leslie, naturalist and artist who illustrated the book, has included a section on field sketching in Chapter 7. The "stake and horizon" method for making a profile across a beach is explained in Chapter 5. Different techniques for measuring waves, tides, and nearshore currents are presented in chapters on waves and tides. The measurements do not require expensive or sophisticated equipment, but they do make it possible to study and record the natural processes that affect the coast.

The examples that demonstrate the principals of coastal oceanography are not limited to a single coast, but are taken from around the world. Norway is used as an example of a fjord coast carved by glaciers and flooded by the sea. Barrier islands are studied along the east coast of North America. The rocky coast examples used vary from New England to the Pacific northwest. The Great Barrier Reef on the Queensland coast of Australia is compared with the West Indies reefs in the Caribbean Sea. Tides are studied from the Bay of Fundy; Immingham, England; Do San, Vietnam; and San Francisco, California.

The book is organized around three main topics: physical processes, coastal landforms, and marine ecology. Physical processes of the coast, including geology, coastal weather and climate, waves, and tides, are covered in the first four chapters. Chapters 5 and 6 discuss origin of beaches, barrier islands, and spits. The organisms and ecology of beaches, dunes, salt marshes, and tide flats are discussed in Chapters 7 and 8. The geology of the rocky coast and ecology of intertidal organisms are included in Chapter 9. The formation of reefs and atolls and the ecology of coral reefs are covered in Chapter 10. It is hoped that this book will become a trusted field companion for those who enjoy walking along the shore, exploring tide pools, or watching birds along the coast.

Acknowledgments

I would like to express my sincere thanks to the many individuals who reviewed the chapters, provided photographs and illustrations, aided in preparation of the manuscript, and supplied encouragement for writing the book. Bud Wobus, Markes Johnson, and Bill Locke of the geology department at Williams College reviewed chapters on coastal geology and marine ecology. Hank Art and Joan Edwards of the biology department at Williams College reviewed chapters on barrier islands and salt marsh ecology. Skip Davis at the University of South Florida and Marty Miller from Houston, Texas, reviewed chapters on waves, tides, and beaches. Ben Labaree and Wendy Wiltse of Mystic Seaport reviewed chapters on marine ecology.

Finally, special thanks must be given to Ann Laliberte and Clare Walker Leslie. Ann spent many extra hours proofreading and typing the manuscript and preparing the lists and captions. Clare drew most of the beautiful illustrations for the book, and also added a naturalist's perspective to several chapters and provided a sounding board for many of the ideas included in the book. My heartfelt thanks to all.

Grateful acknowledgment is given to the following for granting permission to reprint material.

Figure 1–3. After Grant M. Gross, *Oceanography—A View of the Earth,* © 1977 by Prentice-Hall, Inc., reprinted by permission of the publisher, after J. F. Dewey, 1972, "Plate Tectonics," *Scientific American,* 226 (5), 56–58.

Figure 1–4. After W. Anikouchine and R. Sternberg, *The World Ocean, An Introduction to Oceanography,* 2nd ed., © 1981 by Prentice-Hall, Inc., reprinted by permission of the publisher.

Figure 1–6. From John B. Weilhaupt, *Explorations of the Oceans,* © 1975 by Macmillan Publishing Co., reprinted by permission of the publisher, after J. Pattullo, in *The Sea,* edited by M. N. Hill, © 1963 by John Wiley and Sons.

Figure 1–7. Based on data of R. F. Flint, from A. N. Strahler, *Physical Geography*, 4th ed., copyright © 1975 by John Wiley and Sons, New York. Used with permission.

Figure 1–9. From John B. Weilhaupt, *Explorations of the Oceans*, © 1975 by Macmillan Publishing Co., reprinted by permission of the publisher, after Fairbridge, "Changing Levels of the Sea," 1960, *Scientific American*, 202 (5) 77.

Figure 1–12. From John B. Weilhaupt, *Explorations of the Oceans*, © 1975 by Macmillan Publishing Co., reprinted by permission of the publisher, after H. Fisk, 1954, *Journal of Sedimentary Petrology*, vol. 24, pp. 76–79.

Figures 2–7, 2–10. From John G. Navarra, *Atmosphere, Weather and Climate*, © 1979 by W. B. Saunders Company, reprinted with permission of the publisher.

Figures 2–9, 2–15, 2–16, 2–19, 2–23. From Frederick K. Lutgens and Edward J. Tarbuck, *The Atmosphere, an Introduction to Meteorology*, pp. 134, 198, 201, 232, 150, © 1979, reprinted by permission of the publisher, Prentice-Hall, Inc., Englewood Cliffs, N.J.

Figures 2–14, 2–17, 2–18. After Frederick K. Lutgens and Edward J. Tarbuck, *The Atmosphere, an Introduction to Meteorology*, pp. 194, 217, 219, © 1979, reprinted by permission of the publisher, Prentice-Hall, Inc., Englewood Cliffs, N.J.

Figure 3–7. After Moskowitz, 1964, *Journal of Geophysical Research*, vol. 69, pp. 5161–5179.

Figure 3–12. After M. Grant Gross, *Oceanography—A View of the Earth*, 3rd ed., p. 227, © 1982, reprinted with permission of the publisher, Prentice-Hall, Inc., Englewood Cliffs, N.J.

Figures 4–9, 4–10, 4–11, 4–12. After Dean, in Ippen, *Estuary and Coastline Hydrodynamics*, © 1966, Institute of Hydrologic Research, Univ. of Iowa—Ruth M. Ippen.

Figure 4–16. After Van Dorn, *Oceanography and Seamanship*, © 1974, reprinted by permission of the publisher, Dodd, Mead and Company, New York.

Figures 4–18, 4–19, 4–20. From Fox, *Journal of Geological Education*, 1969, vol. 28, pp. 127–129.

Figure 6–3. After D. J. Johnson, 1919, *Shore Processes and Shoreline Development*, © John Wiley and Sons.

Figure 6–4. After J. Hoyt, 1967, *Geological Society of America Bulletin*, vol. 78, pp. 1125–1136.

Figure 6–11. Modified from Fisher, J.J., in Stanley, 1969, with permission of the publisher, American Geological Institute.

Figure 10–19. After Emery, Tracey and Ladd, 1954, "Geology of Bikini and nearby atolls," Geological Survey Professional Paper 260–A.

Figure 10–22. After W. G. H. Maxwell, *Atlas of the Great Barrier Reef*, © 1968 by Elsevier Scientific Publishing Company, Amsterdam, reprinted by permission of the publisher.

Quotation from *The Edge of the Sea* by Rachel Carson, copyright © 1955 by Rachel Carson. Reprinted by permission of Houghton Mifflin Company.

The edge of the sea is a strange and beautiful place. All through the long history of Earth it has been an area of unrest where waves have broken heavily against the land, where the tides have pressed forward over the continents, receded, and then returned. For no two successive days is the shore line precisely the same. Not only do the tides advance and retreat in their eternal rhythms, but the level of the sea itself is never at rest. It rises or falls as the glaciers melt or grow, as the floor of the deep ocean basins shifts under its increasing load of sediments, or as the earth's crust along the continental margins warps up or down in adjustment to strain and tension. Today a little more land may belong to the sea, tomorrow a little less. Always the edge of the sea remains an elusive and indefinable boundary.

Rachel Carson, The Edge of the Sea

Along the Coast of Maine and two common ferns

ONE

Introduction to the coast

Impressions of the Coast

Most people who have lived along a coast or spent summer vacations at the shore have a favorite time or place along the edge of the sea. For some, it is a quiet walk along a deserted beach at sunset with waves lapping gently against the sand. For others, it is a hike along a rocky coast with waves breaking against the base of a cliff (Figure 1–1). Coastal naturalists enjoy scrambling over the rocks at low spring tide in search of barnacles, mussels, sea anemones, and starfish. The character and mood of the coast are constantly changing with storms and tides.

Although each of us has his or her own impression of what a coast should look like, the coasts of the world vary widely from one place to another. The windswept dunes on the outer banks of North Carolina are quite different from the fog-shrouded beaches on the coast of Oregon. The coral reefs exposed at low tide on the Florida coast are very different from the broad expanses of tropical coral reefs along the northeast coast of Australia. In this book, we explore the natural processes and coastal features that are responsible for the different coasts around the world.

FIGURE 1–1.
Wilson's Promontory, a vast natural area on the southernmost tip of Australia.
Courtesy Australian Information Service.

Coastal Geologic Processes

To understand how a coast has evolved and why it has certain features, it is necessary to delve into the geologic processes that formed the coast. This chapter starts with the interior of the earth, and moves on to plate tectonics, which attempts to explain the origin and movement of continents and ocean basins. The crust of the earth has been broken into several large plates that contain pieces of continents and parts of the ocean floor. Major features of many coasts are determined by the position of a continent on a moving plate. For example, the Atlantic and

Pacific coasts of the United States have vastly different characteristics. The Pacific coast is relatively straight with steep cliffs and broad beaches. The Atlantic and Gulf coasts, on the other hand, are indented with bays and estuaries and draped with a chain of barrier islands from Massachusetts to Texas.

The position of the coastline along the edge of a continent also changes with rise and fall of sea level. During the past few million years, sea level has risen and fallen several times as glaciers have advanced and retreated over the continents. About 20,000 years ago, glaciers covered about a third of North America. When the glaciers partially melted, sea level rose about 130 meters. About 3000 years ago, sea level had almost reached its present level, but it is continuing to rise about 20 to 30 centimeters per century. As sea level rises, coastal areas are flooded and the shoreline retreats toward the land.

Earthquakes and faulting have caused portions of the land to rise or fall relative to the sea. In some areas, ancient shorelines have been lifted several hundred meters above the sea, and in other areas, coastal features have been dropped well below sea level.

Geologic agents including glaciers and rivers, which are important in shaping the land on the continents, also have an influence on the coasts. Rivers and glaciers carve deep valleys that are flooded by a rise in sea level forming bays and estuaries. The sediment load carried by the rivers and glaciers is eventually dumped in the sea, forming deltas and glacial deposits. These sediments are reworked by waves and currents, which supply much of the sand to the beaches.

Waves, tides, and longshore currents are responsible for the day-to-day changes in the character of the coast. Waves are constantly eroding the cliffs and piling sand on beaches. Tides raise and lower sea level twice each day so that the effects of waves are spread over a wider area of the beach. Longshore currents move sand along the shore, building sand spits and filling in harbors.

Finally, organisms that live along the coast contribute to the form and shape of the shore. In temperate and polar regions, organisms such as barnacles and mussels live on the rocks, but the shape of coast is determined by waves and currents. In the tropics, organisms build large reefs that dominate the coast. Many of the islands of the Pacific are coral reefs or atolls built on long-extinct volcanoes.

Interior of the earth

If you could slice the earth in half with a large knife, the interior of the earth would resemble a giant peach (Figure 1–2). The three major zones in the earth—core, mantle, and crust—correspond to the pit, flesh, and skin of a peach. However, the zones within the earth form concentric spheres like shells of an onion. The compositions and thicknesses of the zones have been determined from the study of earthquake waves that have passed through the earth.

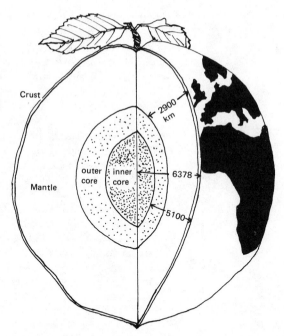

FIGURE 1–2. The crust, mantle, and core of the earth resemble the skin, fruit, and pit of a peach.
Drawing by C. W. L.

The core of the earth is about 7000 kilometers (about 4320 miles) in diameter and occupies about 31.5 percent of the earth's volume. The core contains two parts, a solid inner core and a liquid outer core. The inner core is composed of iron with small amounts of nickel, has a diameter of about 2432 kilometers, and is about 4800 degrees Centigrade. The outer core is thought to be liquid metallic iron with some carbon, silicon, and sulfur. The average temperature of the outer core is estimated to be about 4000 degrees Centigrade. Slow movements in the liquid outer core are probably responsible for the earth's magnetic field and may also be the source of the heat that sets up convection currents in the mantle.

The thick shell of rocky material that surrounds the core is known as the mantle. The mantle is about 2800 kilometers thick and contains almost 68.1 percent of the earth's volume. Although the mantle is solid, it is under extreme heat and pressure, and sometimes it flows like a highly viscous fluid such as molasses. The outer 700 kilometers, which flow slowly under pressure, are known as the *asthenosphere*.

A rigid sphere known as the *lithosphere* overlies the asthenosphere. The lithosphere contains two layers, an upper layer that is the crust, and a lower layer formed from the upper part of the mantle. On the continents, the crust is 25 to 40 kilometers thick, reaching 60 kilometers under some mountain ranges. The bulk of the continental crust is granite that is overlain by a thick veneer of metamorphic and sedimentary rocks. Under the floor of the oceans, the crust consists of a layer of basalt about 5 kilometers thick that is usually covered by a layer of sediment up to about 1 kilometer thick. The rocks of the continental crust are lighter and thicker than the rocks of the oceanic crust and float above the level of the ocean like ice cubes bobbing in a glass of water. Sea water fills in the low places between the continents, forming the ocean basins.

Plate tectonics   The concept of plate tectonics evolved in the late 1960s and early 70s to explain the present arrangement of continents and ocean basins. According to the hypothesis, the lithosphere, which includes the earth's crust and the upper part of the mantle, is segmented into several large plates that fit together like a jig-saw puzzle (Figure 1–3). The rigid lithospheric plates are moving slowly on the viscous athenosphere. The plates are moving away from zones of spreading where new oceanic crust is created. The zones of spreading are called divergent plate boundaries. Crustal material is consumed where the plates come together at convergent plate boundaries. The continents ride along on top of the plates like cargo on top of barges.

The difference in temperature between the earth's core and mantle provides the driving force that moves the lithospheric plates. Early in the earth's history, the temperature rose in the interior of the earth and the heavier elements, including nickel and iron, descended toward the center to form the core. The lighter elements, including silica and aluminum, rose toward the surface and cooled to form the mantle and the crust.

As the earth cooled, heat escaped from the interior by conduction and convection. Conduction is the slow transfer of heat by ion-to-ion or molecule-to-molecule impacts. The hotter, faster-moving ions or molecules cause the slower-moving ions or molecules to move faster. Metals are good conductors of heat. Therefore, when a spoon is placed in a hot cup of coffee, the heat in the coffee sets the molecules in the spoon in motion. Heat is transferred from the coffee to the spoon and eventually to the air. Rocks within the earth are poor conductors of heat. Therefore, it took an extremely long time for the accumulated heat to pass out of the earth by conduction. The amount of heat moving out of the earth's interior by conduction is quite small relative to the heat energy radiated from the earth's surface. Light energy from the sun is absorbed by the earth and radiated as heat energy to heat the atmosphere.

Convection is the motion of a fluid caused by uneven density related to heat. A hotter, less-dense fluid rises, whereas a cooler, denser fluid sinks. In convection, heat is carried along with the moving fluid. When the bottom of a pot of coffee is heated, the hot water expands and rises while the cooler water descends, forming a convection cell. At the base of the earth's mantle, heat from the core forms hot areas that generate large convection cells within the mantle.

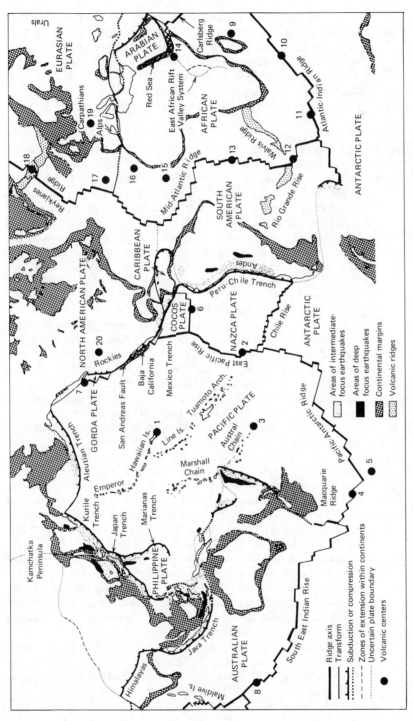

Legend:

1. Hawaii
2. Easter Island
3. Macdonald Seamount
4. Bellany Island
5. Mt. Erebus (Antarctica)
6. Galapagos Islands
7. Cobb Seamount
8. Amsterdam Island
9. Reunion Island
10. Prince Edward Island
11. Bouvet Island
12. Tristan de Cunha
13. St. Helena
14. Afar
15. Cape Verde
16. Canary Islands
17. Azores
18. Iceland
19. Eifel
20. Yellowstone

Ridge axis
Transform
Subduction or compression
Zones of extension within continents
Uncertain plate boundary
Volcanic centers

Areas of intermediate-focus earthquakes
Areas of deep focus earthquakes
Continental margins
Volcanic ridges

FIGURE 1–3. Worldwide lithospheric plate system, showing major centers of volcanic activity.

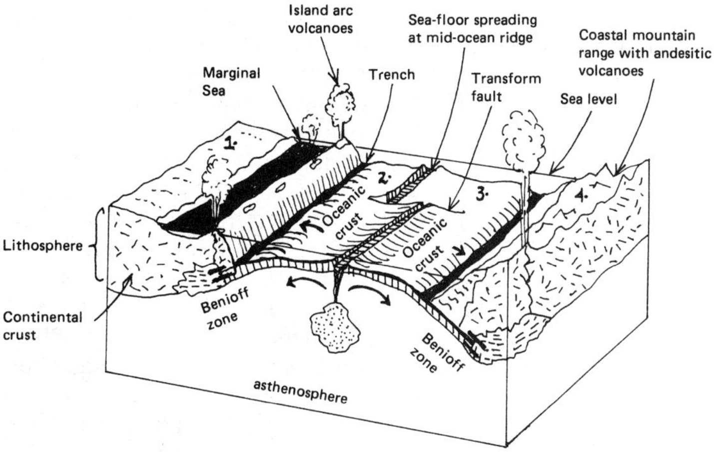

FIGURE 1–4.
The processes of the formation of new crustal plates at spreading centers, sea floor spreading, and destruction of plates by subduction at trenches. Each different plate is numbered.

Convection cells within the mantle rise and spread out along the base of the lithosphere (Figure 1–4). As rising convection cells diverge near the earth's surface, rigid lithospheric plates are carried along on the top of the convection cells. The top of the convection cell acts like a conveyor belt that carries the lithospheric plate away from the spreading center.

Heat from convection currents causes the crust along the spreading boundaries to rise, forming ridges. A network of ridges 60,000 kilometers long extends from the middle of the Atlantic Ocean, around Antarctica, to the east Pacific off western North and South America and into the Indian Ocean (Figure 1–3). A rift valley often is found along the crest of the ridge. Molten rock called magma moves up from below along the spreading boundary. Where the magma breaks through to the surface, volcanoes form along the ridge crest. Surtsey, a new volcanic island, appeared off the south coast of Iceland on November 14, 1963. Iceland is located along the ridge crest in the middle of the Atlantic Ocean.

The heated ridge is about 4 kilometers higher than the surrounding ocean floor, and its flanks gradually slide away from the rift valley. As oceanic crust spread away from the ridge, forming the ocean floor, it slowly cooled, became more dense, and settled back to its original depth below the surface of the ocean.

New material is added to the edges of the plates along spreading or diverging boundaries as the oceans are expanding. Each side of the Atlantic is spreading away from the mid-oceanic ridge at about 2 centimeters per year, giving a combined spreading rate of about 4 centimeters per year. Therefore, Europe and North America are drifting apart about 1 meter every 25 years. The distance between New York and Paris grows a kilometer longer every 25,000 years, or about a mile every 40,000 years. The two sides of the Atlantic have been spreading for about 150 million years and are now about 6,000 kilometers, or 3,750 miles, apart.

Where two plates move toward each other, or converge, one plate may descend beneath the other along a subduction zone (Figure 1–4). As a plate moves slowly from a ridge to a subduction zone, thin layers of sediment accumulate on the top of the plate. This material consists of mud, silt, and fine sand carried out from the continents, and ooze composed of tiny shells of planktonic organisms that lived in the upper layers of the ocean. As a plate descends along a subduction zone, some layers of sediment are scraped off the top and plastered against the edge of the continent. The rest of the sediment is carried down with the descending plate, where it is melted and rises again to form continental volcanoes or island arcs. Mount Saint Helens, which erupted violently in 1980, is an example of a continental volcano above a descending plate. Along the Pacific coast of North America, the Gorda plate is descending beneath the North American plate, producing the volcanoes of the Cascade Range in Washington and Oregon (Figure 1–3). Japan and the Aleutian Islands in the northwest Pacific are examples of island arcs that lie adjacent to deep oceanic trenches. The subduction zones bordering the Pacific Ocean form a "ring of fire" where many of the major earthquakes and volcanoes of the world occur.

Coastal tectonics

Many of the major features of a coastline are the direct result of plate tectonics. The horizontal movement of large lithospheric plates produces vertical movements along the coast that are known as coastal tectonics. A continent is embedded within a plate that moves slowly across the surface of the earth. The edge of a continent facing toward a divergent plate boundary or spreading zone is referred to as a trailing-edge coast. Where the edge of a continent lies adjacent to a converging plate boundary or subduction zone, it is known as a leading-edge coast. Where two plates are moving past one another without spreading or subduction, they form a sliding-edge coast.

A trailing-edge coast is characterized by a wide continental shelf with a gentle slope. When a continent splits to form an ocean basin along a divergent plate boundary, hot magma rises through the crust forming a mid-oceanic ridge. As the lithospheric plates spread apart, the ocean floor slowly descends to form the abyssal plain. The edge of the continent is also bent downward by the weight of the adjacent oceanic crust. Rivers draining the interior of the continent dump a thick wedge of sediment along the coast. The waves and currents redistribute the sediment along the shore, forming beaches, spits, and barrier islands. As the ocean continues to expand, the sediment along the coast accumulates and the wedge of sediment increases in width and depth.

Along the southeastern coast of United States, the coastal plain sediments are more than 300 kilometers wide and 10 kilometers thick. The western edge of the coastal plain is marked by the Appalachian Mountains, with the eastern edge extending seaward to the boundary between the continental shelf and slope at a depth of about 200 meters. The slope of the upper surface of the coastal plain is about 1 to 1000, or about 1 meter vertical drop per kilometer.

Rivers that flow over the coastal plain were drowned to form large bays including Chesapeake and Delaware Bays. Barrier islands extend along the east coast of United States from Massachusetts to Florida, and along the Gulf coast to Texas. In New England and the southeast coast of Canada, the trailing-edge coast has been modified by glacial erosion and deposition.

The trailing-edge coasts of Australia, Africa, and parts of South America have been very stable with little uplift or downwarp since the last ice ages. The continents of Australia and Africa are embedded within plates and do not have adjacent subduction zones. The east and west coasts of Africa face toward spreading centers in the Atlantic and Indian Oceans (Figure 1–3). South America has a leading-edge coast on the west and a trailing-edge coast on the east.

Leading-edge coasts form along convergent plate boundaries where oceanic crust is thrust beneath a continent along a subduction zone. A deep trench forms where oceanic crust is bent downward into a subduction zone. The edge of a continent along the subduction zone is folded, faulted, and uplifted by the confining pressure of the converging plates. Leading-edge coasts are relatively straight with steep cliffs and narrow continental shelves.

The rugged and spectacular coast of Washington and Oregon provides an excellent example of a leading-edge coast. The high-straight cliffs provide long vistas to the north and south along the coast. Along many sections of the coast, a wave-cut platform extends along the base of the cliffs. During the summer, the rock platform is covered by a thin veneer of sand, forming a wide, flat beach (Figure 1–5). Much of the sand is removed by the fierce winter storms, and the rock pavement is exposed at low tide.

The coast of California is strongly influenced by the San Andreas fault. The fault heads south along the coast from San Francisco to Los Angeles. The portion of the coast to the west of the fault is moving slowly to the north relative to the rest of the state. Bodega Bay and other coastal features were formed directly by the San Andreas fault. Since the plates are moving past each other without spreading or subduction, it would be considered a sliding-edge coast. A horizontal fault where two plates slide past one another is also called a transform fault.

The west coast of South America is a leading-edge coast with the Peru Trench to the west and the Andes Mountains parallel to the coast. The narrow coastal plain is confined by the trench, marking the subduction zone and the mountains along the uplifted edge of the continent.

FIGURE 1–5. Sand blowing along a windswept beach at
Newport, Oregon.
Photo by W. T. Fox.

The main fabric or character of a coast is determined by
plate tectonics, with the differences along a coast accounted for
by other geologic processes, including glaciers, waves, tides, and
currents. The major differences in character between the Atlantic and Pacific coasts of the United States can be accounted for
by coastal tectonics. The Pacific coast is on the leading-edge of
the continent facing a subduction zone to the west. The Atlantic
coast is a trailing-edge coast that has spread away from the
mid-Atlantic ridge. The distribution of bays, headlands, and
beaches along each coast is a product of the local geologic
agents that are superimposed on top of the tectonic imprint.

*Fluctuations
in sea level*

Many features we see along a coast have been affected by
changes in sea level. We often think of sea level as being constant, but actually it is in a constant state of change. Contour
maps used to locate places on land are based on average sea
level, with contour lines given in feet or meters above or below
mean sea level. However, sea level constantly changes, either
rapidly due to waves, tides, and storms, or slowly due to the advance and retreat of glaciers or the rise and fall of land areas.

The relative changes in sea level can be grouped into four major categories: local, seasonal, regional, and eustatic. Local sea level changes affect a small section of coast for a short interval of time and cause significant damage to man-made structures, but they do not leave a lasting impression on natural coastal features. Seasonal changes in sea level cover a wide area, but are usually quite small and do not have an impact on the coast. Regional uplift or downwarp of the coast influences a large area for a long time. Worldwide or eustatic sea level changes affect the entire ocean for hundreds or thousands of years. Regional and eustatic changes in sea level happen very slowly and may not be noticed in our day-to-day lives, but they have a large influence on the shape and evolution of the coastlines.

Local sea level changes. Local rise and fall in sea level may last from a few seconds to several months, but include only a small portion of the coast such as a bay or beach, and cover a few hundred square kilometers of the sea surface. Waves and tides produce the most obvious local changes in sea level and will be considered in greater depth in following chapters. Waves produce instantaneous changes in sea level, from a few centimeters up to 30 meters, but do not affect the average sea level. Although it is not possible to predict rise of sea level due to a single wave, the average wave height is predictable from the wind speed and direction during a storm. Tides produce a twice-daily rhythmic rise and fall in sea level, which can be accurately predicted for several years in advance. Tides resemble wind waves in that they produce an oscillation about the average sea level. Although the tides are produced by the gravitational attraction of the sun and the moon, the tidal range and the times of high and low tide are controlled by the size and shape of the ocean basin. Waves and local weather conditions can alter the time and range of the tides along a coast.

When wind blows in an onshore direction for several hours during a storm or hurricane, the water is pushed toward the coast, resulting in a storm surge. In a hurricane much of the coastal damage comes from flooding by a storm surge. If large waves accompany the storm surge at high tide, coastal damage can be widespread. Since sea level returns to normal soon after a storm subsides, the change is short-lived and most of the damage is to man-made structures. However, dunes and cliffs beyond the reach of normal waves are often breached by waves when a storm surge raises sea level.

Local sea level changes are also produced by a rapid change in barometric pressure. When a storm or frontal system passes over a coastal area, it is often accompanied by a rapid rise or fall in barometric pressure. When barometric pressure drops near the center of a storm, the column of air directly overhead weighs less than it did before. The decrease in barometric pressure produces a corresponding small rise in sea level.

The greatest effects of barometric pressure changes can be seen in enclosed bodies of water such as lakes or land-locked seas. When a cold front passes over the coast, there is often a rapid rise in pressure as the heavy, cold air moves into an area. The sudden increase in pressure causes a drop in water level that can set up a standing wave in a lake. In a standing wave, the water oscillates vertically between fixed points, called nodes, without progression. The points of maximum rise and fall are called antinodes. A standing wave on a lake is called a seiche. The water continues to rise or fall along the shore at about 15-minute to half-hour intervals until the disturbance subsides.

Seasonal changes in sea level. Some changes in sea level are annual and are caused by seasonal changes in atmospheric pressure. The seasonal changes in sea level vary from about 10 to 20 centimeters, but extend over large areas of the oceans (Figure 1–6). During winter in the northern hemisphere, cold, dense arctic air masses invade the mid-latitudes and cause a widespread lowering of sea level. Due to strong onshore winds, however, sea level off the coasts of Washington and Oregon is raised during the winter months. During the summer, the cold air is replaced by warm tropical air and sea level in the Northern Hemisphere is above normal. In the Southern Hemisphere, the sea level is slightly higher during the summer months of February, March, and April. During winter in the Southern Hemisphere, cold air masses spread out from the Antarctic. Advancing cold air lowers sea level in a belt extending east and west from Australia.

Although many of the seasonal changes in sea level in the higher latitudes are due to increased barometric pressure, the rise in sea level in the temperate regions may be due to seasonal heating of the oceans. When temperature is increased, water expands slightly and the surface of the ocean rises. The water that was heated by the sun is mixed through the upper few hundred meters of the ocean. Thermal expansion is limited in extent but is up to a meter in height in the middle of the Pacific Ocean. Although seasonal changes do not affect coastal erosion, they do have an important influence on major ocean currents, including the Gulf Stream.

Coastal uplift or downwarp. Regional sea level changes take place when portions of the earth's crust move up or down. The area of large vertical movements due to warping of the crust usually exceeds 10,000 square kilometers (4000 square miles). Although vertical movement may be hundreds or even thousands of meters, it generally does not affect an entire continent. The vertical movement is caused by the advance or retreat of large glaciers or by the collision or separation of continents on the margins of lithospheric plates.

During the most recent advance of the glaciers about 20,000 years ago, large portions of Europe and North America

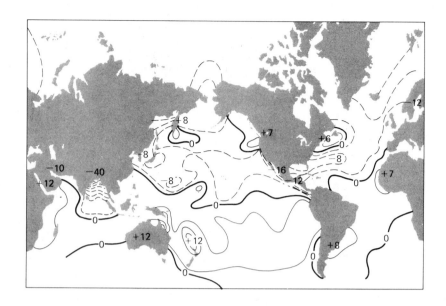

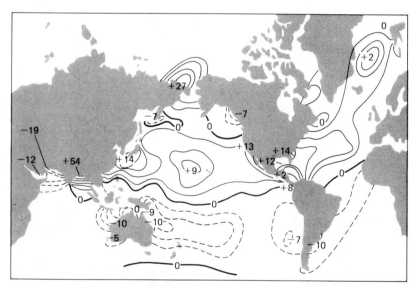

FIGURE 1–6. A. World mean sea level deviations averaged for February, March, and April; and B. World mean sea level deviations averaged for August, September, and October.

were covered by a thick mantle of ice. The weight of the ice on the continents pushed the base of the lithosphere down into the asthenosphere. When the glaciers melted, the lighter continents slowly rose in a process called isostatic rebound. If you hold a cork under the water, the cork will bob up when you release your hand. Similarly, a continent will rise when the weight of the glacier is removed.

In North America, the southern boundary of glacial ice extended along a line from Cape Cod and Long Island across the Ohio Valley to Kansas and Nebraska (Figure 1–7). Since the ice started to melt along its southern margin, the southern part of the area covered by the glacier has rebounded more than the north. The northern ends of the Great Lakes are presently rising more rapidly and the southern shores are being flooded.

Regional changes in sea level related to horizontal movement of crustal plates usually extend over a longer period of time, but can occur quite rapidly. Raised coastal terraces on the Oregon coast are evidence of recent uplift (Figure 1–8). Portions of the California coast have been uplifted several hundred meters over the past million years. In the Indian Ocean where India collided with Asia, the lithospheric plate motions are still producing the Himalayan mountain range.

Eustatic sea level. When the sea level changes uniformly over the entire world, it is referred to as an eustatic sea level change. Although the eustatic changes in sea level occur over a much wider extent, the vertical rise or fall in sea level is less than regional changes. The maximum eustatic sea level change would be about 200 meters (650 feet) between the time of the full extent of glaciation and total melting.

Eustatic sea level changes are generally due to a change in the volume of water in the ocean or a change in the size of the

FIGURE 1–7.
Maximum extent of Pleistocene ice sheets.

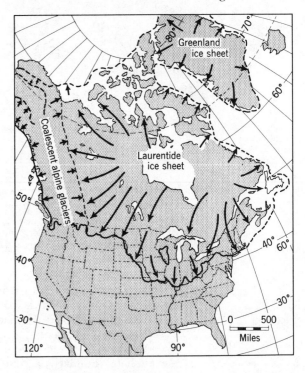

FIGURE 1–8. Wave-cut terraces were raised above sea level on
the Oregon coast.
Photo by W. T. Fox.

ocean basins. A change in the volume of the ocean can be accomplished by converting liquid water into water vapor by evaporation, or by changing liquid water to ice by freezing. Water is constantly being evaporated from the surface of the ocean in the tropics and transported toward the poles in the form of clouds. Most of the water falls as rain over the ocean. However, some of the water is carried over the continents and falls as rain or snow. Sea level does not change rapidly because a balance is maintained between the water removed from the ocean by precipitation that falls on the land, and the water returned to the sea by rivers.

When the climate on the earth turned colder and more snow fell in the upper latitudes, large glaciers accumulated on the continents. As the glaciers increased in size, the water that evaporated from the surface of the ocean fell on the glacier as snow and was transformed into ice within the glacier. When glaciers reached the edge of the continent, they launched icebergs that flowed in the ocean current and eventually melted in warmer water. While the glacier is still on the continent, ice melts around the margin of the glacier, forming large lakes or rivers that return the water to the ocean. When the rate of melting at the margin of the glacier exceeded the rate of accumulation of new ice by means of snow, the glacier decreased in size and the level of the oceans rose. The water that was frozen on the sea surface does not affect sea level, because the floating ice displaces an equal weight of sea water and the ocean maintains a constant level.

During the recent major advance of the glaciers, the Wisconsin glacial period, which extended from 10,000 to 70,000 years before present, sea level was about 100 meters (330 feet) below present sea level (Figure 1–9). It has been estimated that about 10 million cubic miles of water were incorporated in the glacial ice during the Wisconsin glaciation, which accounts for the 100-meter drop in sea level. During the Wisconsin glaciation, the coastline on the eastern seaboard of the United States was about halfway out to the edge of the continental shelf. New York City was an island in the Hudson River about 160 kilometers inland from the coast. During the Illinoian Glacial advance, extending between 90,000 and 130,000 years before present, sea level dropped about 85 meters.

During the Sangamon Interglacial period between the Illinoian and Wisconsin Glacial intervals, sea level was about 18 meters above present sea level. There were still ice caps in the Arctic and Antarctic regions, but major ice caps had retreated from the Canadian Arctic. Beaches and sea cliffs preserved at higher levels represent successive times when sea level stood still during the Sangamon Interglacial. The Surrey Strandline, which is the highest sea level, is about 30 meters above present sea level. Several ancient shorelines are also present as concentric ridges around the southern ends of the Great Lakes, marking the earlier extent of the lakes. As the land tilted northward due to

FIGURE 1–9. Glacial sea level decline after the Pleistocene.

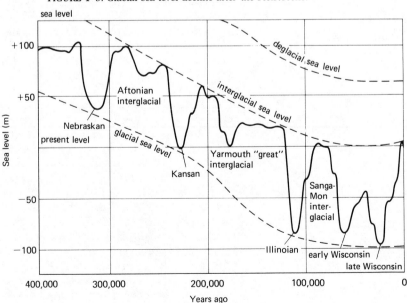

isostatic rebound, the lakes retreated to the north, leaving a series of shorelines stretched across the countryside.

Evidence for sea level at lower elevations than today must be looked for on the bottom of the ocean. On the continental shelf off eastern United States, a step on the ocean floor that has been traced 185 miles may represent a submerged sea cliff. Submarine canyons carved into bedrock at 60 meters below sea level can be traced into rivers on land. These submerged valleys were carved by rivers when sea level was lower during a glacial period.

If the present ice caps on Greenland and Antarctica were to melt, sea level would rise about 70 meters. The Port of New York would be submerged estuary with docking facilities around Albany, New York. Manhattan Island would be represented by a few heliports on top of the highest skyscrapers.

Coastal Types
Primary coasts

Although large-scale landforms on a coast such as mountain ranges and coastal plains are determined by the horizontal movement of lithospheric plates, smaller features are controlled by coastal erosion and deposition. The edge of the continent is sculptured and molded by rivers, glaciers, and earthquakes. The initial coast that inherited many of the original characteristics of the landscape is called a primary coast.

After the coast has been submerged, waves and currents begin to affect the coast by eroding the landscape and depositing sediment along the coast. In the tropics, organisms along the coast build reefs that overshadow the processes of erosion and deposition. A coast that has been modified by marine erosion or deposition, or by the action of organisms, is considered a secondary coast.

To understand coastal features, it is necessary first to study the geologic processes that are operating on land, and to examine the types of landscapes they produce. From the initial landscape, we move on to the primary coastal features formed when different types of landscapes are flooded by the sea. Finally we will consider the secondary coastal features that are brought about by the action of waves, tides, currents, and organisms.

Ria coasts. The characteristics of the landscape are caused by uplift followed by erosion. Rain that falls on the land soaks into the ground until the pore spaces in the soil are filled, then starts to run off on the ground surface. At first, water flows uniformly over the surface of the land as sheetflow, but eventually it gathers into small channels that merge into rivers and flow to the sea. When water flows across the land, it picks up loose particles of soil and rock and carries them into the streams. The streams meander across the countryside, carrying sand, silt, and clay that are eventually deposited in the ocean.

When a river is flooded by the sea, it forms a broad estuary, known as a Ria coast. Ría is the Spanish word for estuary, which is the drowned mouth of a river. The Ria coastline resembles a maple leaf with arms of the sea extending inland into bays or estuaries. The bays of a Ria coast are usually wide and deep near their mouth, becoming narrow and shallow as they extend inland, eventually merging with rivers that still flow toward the sea. Chesapeake Bay on the east coast of United States provides an excellent example of a Ria coast. The lower part of the Susquehanna River flowing through eastern Pennsylvania was drowned by a postglacial rise in sea level to form Chesapeake Bay.

Glacial coasts. In the mountains where snowfall in the winter exceeds summer melting, large snow fields accumulate. When the snow becomes deep enough, it turns to ice and forms a glacier. As a glacier flows downhill following old river valleys, it carves a *U*-shaped trough with steep walls and a relatively flat bottom. Yosemite Valley in California is a good example of a *U*-shaped valley cut by a glacier.

Where glaciers reach the coast, an advancing glacier may actually extend into the ocean and erode its valley beneath the sea. When a glacier advances into the sea, its snout is attacked by waves and melting takes place along the cracks within the ice. Large icebergs are calved off the front of the glacier and float with ocean currents. As the glacier advances, it continuously melts while new ice is being added from the snowfield. The glacier will continue to move forward as long as new ice is added, but the front of the glacier will retreat if the melting and calving of icebergs exceeds the accumulation of new ice.

As the front of the ice retreats up the *U*-shaped valley, the sea floods the valley and a fjord is formed (Figure 1–10). Although a fjord may be formed when a glacial valley is flooded by a rise in sea level, submergence is not essential because the glaciers can erode below sea level. Fjords with nearly vertical walls, hanging valleys, and deep harbors are typical of the glaciated coasts of Norway, Alaska, Labrador, and New Zealand.

The coast of Maine has been modified by a continental ice sheet that gouged broad glacial troughs, creating long bays and rounded headlands. Somes Sound in Mount Desert Island is a fjord formed by the submergence of the Maine coast. The straight-line distance between the northern and southern boundaries of Maine from Canada to New Hampshire is only 375 kilometers (225 miles), but the coastline of the bays, islands, and promontories measures about 4200 kilometers (2500 miles).

When a glacier moves across a continent, sediment is carried within and on top of the glacier. When the forward motion of the glacier stops, sediment is deposited along the edge of the ice as a moraine. Streams flowing from the glacier deposit sand

FIGURE 1–10. Sognefjord, a fjord on the Norwegian coast carved by rivers, polished by glaciers, and flooded by the ocean.
Courtesy of the Norwegian Travel Association.

and silt in braided channels, forming an outwash plain. The north shores of Cape Cod, Nantucket, Martha's Vineyard, and Long Island are flooded moraines, and the south shores are submerged outwash plains. Other landforms including drumlins and ground moraines are also created beneath the ice as it moves across the countryside. The islands in Boston Harbor are submerged drumlins with their long axes parallel to the flow direction of the ice.

Landslide coasts. Landslides, which are common along the California and Oregon coasts, form lobe-shaped features that extend into the surf zone. Landslides usually contain loose and unconsolidated material, and the toe of the landslide is rapidly eroded by wave action (Figure 1–11). When the toe is removed, the landslide moves again, and new landslides form farther inland. The large landslide at Portuguese Bend south of Los Angeles, California, is an active landslide coast. The toe of the landslide is removed by waves as the slide moves toward the sea.

FIGURE 1–11. Waves eroding the base of a cliff cause a landslide on the Oregon coast.
Photo by W. T. Fox.

Karst shorelines. When layers of limestone extend above the water table, weak acids from the overlying soil etch the limestone and eventually form caves. As the cave expands, the roof may collapse and a sinkhole develops. An area that is pockmarked by sinkholes is known as a karst topography. When a sinkhole is submerged beneath the sea, a very irregular coastline develops, which is known as a karst shoreline. The coast of Barbados in the Caribbean has a karst shoreline eroded into old caves that formed in solution cavities in reef limestones.

Delta coasts. When large rivers enter the ocean, sediment is dumped at the mouth of the river, forming a delta. If the river is carrying a relatively small amount of sediment, a rise in sea level will form a drowned river valley or Ria coast. However, when the river is draining a broad area and carrying a large volume of sediment, the river channel builds out into the ocean on top of its own deposits. The delta coastline is a delicate balance between the sediment load deposited by the river, the erosion of waves, and the sweeping action of longshore and tidal currents.

When the river dominates, a bird-foot delta is formed. The Mississippi Delta near New Orleans is an excellent example of a bird-foot delta (Figure 1–12). Natural levees are built up on each side of the main channel when the Mississippi overflows its banks and spreads onto the floodplain. Sediment is deposited in front of the mouth in the form of a lunate bar, thus extending the length of the main channel into the Gulf of Mexico. When the sediment in the lunate bar clogs the channel, the channel splits into distributary channels, forming the claws typical of the bird-foot delta.

When waves and longshore currents dominate the river, the sediment is spread along the coast to form an arcuate or cuspate delta. The Nile Delta in Egypt is typical of an arcuate delta where the river has one dominant channel and several secondary channels that were active during floods. Since construction of the Aswan Dam, the Nile Delta is no longer subjected to annual flooding. The silt and sand that replenished the delta plain and were referred to as the "gift of the Nile" are now trapped behind the Aswan Dam.

Where the tide is important along the shore, the mouth of the estuary expands seaward like a trumpet with a series of long sand bars in the channel. Instead of forming a delta, the bars

FIGURE 1–12. A bird-foot delta was formed from sand and silt deposited at the mouth of the Mississippi River.

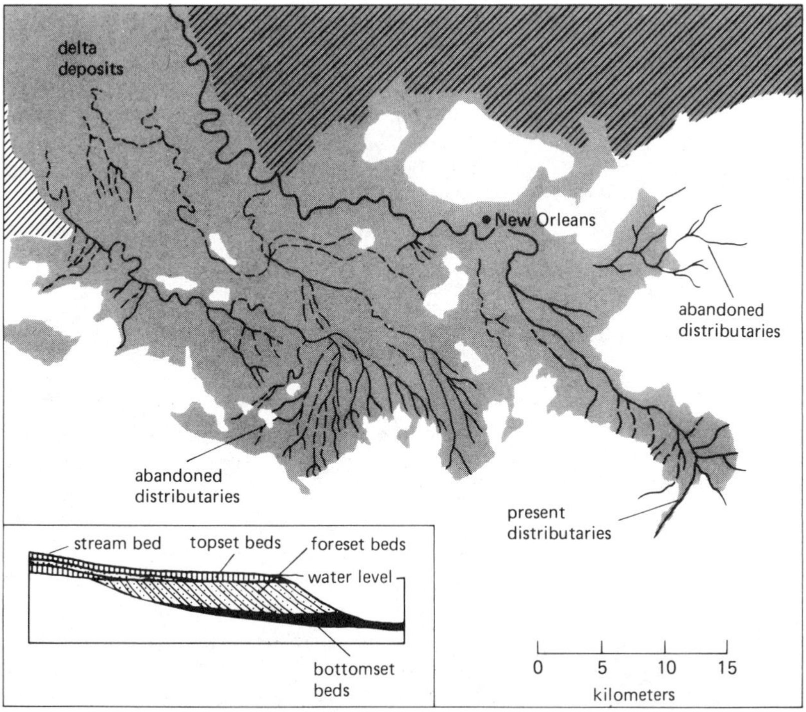

are deposited by the strong ebb and flood currents moving in and out of the bay. The Ord Delta in Australia and the Klang Delta in Malaya are examples of tide-dominated deltas.

Volcanic coasts. A volcanic coast is formed where a volcano erupts along the coast or forms an island in the sea. There are two major types of volcanoes, the andesitic volcano, which is highly explosive and forms a steep-sided cone, and the basaltic volcano, which is relatively quiet and produces a broad, flat shield volcano. Mount Saint Helens is a typical andesitic volcano that formed directly over the subduction zone of a convergent plate boundary. The cone is composed of loose volcanic material interbedded with layers of lava. In some instances, an andesitic volcano will explode violently and collapse into the crater, forming a caldera. In some caldera eruptions, several cubic kilometers of rock are blasted into the atmosphere. Some of the rock falls back into the crater, forming the floor of the caldera. When a caldera explodes in a volcanic island, the inside walls form steep cliffs facing toward the center of the extinct volcano. The Island of Thira, earlier known as Santorina, has steep cliffs that surround the center of a collapsed caldera. When the volcano on Santorina exploded, the ash may have been responsible for the downfall of the Minoan civilization on the Island of Crete.

Volcanoes that erupt in the ocean are composed of basalt, the material that forms the ocean floor beneath the sediments. Basalt erupts quietly as lava that flows down the sides of the volcano. The Hawaiian Islands are examples of large shield volcanoes with gently sloping flanks. Mauna Kea, the highest mountain in Hawaii stands 4206 meters (13,796 feet) above sea level and more than 10,670 meters (6.6 miles) above the floor of the ocean. Most of the coast of Hawaii consists of recent lava flows that show little evidence of erosion, except along the northeast coast, from Hilo to the north end of the island. The older islands in the Hawaiian chain are extinct volcanoes that show considerable evidence of both stream and wave erosion. The older volcanoes are slowly sinking into the floor of the ocean, subsiding under their own tremendous weight. The stream valleys surrounding the dormant volcanoes are flooded as the volcanoes sink, developing a Ria coastline with a radial pattern. When volcanic activity ceases, corals rapidly colonize the flanks of the volcanoes and form reefs. As the volcanoes sink beneath the sea, the coral reefs continue their upward growth and form coral atolls.

Secondary coasts   Once a primary coast has been formed by submergence or emergence, marine processes begin to transform it into the secondary coast. The secondary coasts are of three major types: rocky coasts, beaches and barrier islands, and reefs. The secondary coasts are dealt with in detail in later chapters, so they will be only briefly considered at this time.

Climate is a major factor in determining whether erosion or deposition will dominate, or whether the coast will be taken over by organisms. Storm waves most frequently occur in the middle latitudes where cold, dry polar air masses come in contact with warm, moist tropical air masses, producing cold and warm fronts. In the Pacific, the storm belt extends in an arc from the coast of Japan through Alaska to the northwestern United States. In the Atlantic, the New England States, the Maritime Provinces of Canada, and northern Europe are in the storm belt. The winter storms are more severe because the contrast between the arctic and tropical air masses is greater.

In the southern Hemisphere, the storm belt is located mainly over the open ocean. Patagonia, at the tip of South America, is the only coastal area that is directly affected by the southern storm belt. However, swells from the Antarctic have been traced northward across the equator to the Aleutian Islands.

Storm coasts. The storm wave environment is a regime of coastal erosion where rugged cliffs and beautiful landforms develop. Storm waves initially cut notches in bedrock cliffs, forming gravel beaches. As erosion continues, the notch undercuts the cliff and landslides occur. As the cliffs are eroded back, an abrasion platform develops along the base of the cliffs. If sea level remains stable for a long interval of time, cliff erosion will continue until the abrasion platform is very wide.

The material that is eroded from the cliffs in the storm wave environment is usually gravel or coarse sand that forms beaches or long spits (Figure 1–13). Where the sea cliffs alternate with bays on a rolling countryside, the sand and gravel are deposited across the bays, forming large bay mouth bars. If the bay is blocked off from the sea by a bar or spit, a lagoon will form behind the bar. Eventually, the lagoon will fill with sediment, forming a salt marsh.

Swell coasts. Large depositional features are commonly formed in the swell environments. The swells that originate in the storm belt travel for thousands of miles along great circle paths. The southwesterly swell is dominant in the Southern Hemisphere, and the northwesterly swell is more important in the Northern Hemisphere. Therefore, coasts along the western edges of continents will have stronger swells that coasts along the eastern edges.

Broad sand beaches and large barrier islands are constructed in the swell wave environment. The barrier islands are long, low islands that are roughly parallel to the coast. The islands are flanked by a sand beach on the ocean side and separated from the mainland by a broad lagoon or salt marsh. Barrier islands are found along many coasts of the world including North and South America, Africa, and Australia (Figure 1–14).

FIGURE 1–13.
Sand from eroding cliffs forms
beaches at Aireys Inlet Coastline,
Victoria, Australia.
*Courtesy of the Australian
Tourist Commission.*

FIGURE 1–14. A barrier island and marsh on the Florida coast
near Saint Augustine.
Photo by W. T. Fox.

Along the east coast of the United States, barrier islands extend from Massachusetts to Florida, and in the Gulf of Mexico along the Texas coast.

Tropical coasts. Although waves, tides, and currents are the dominant forces that form coasts in the northern latitudes, organisms take over that role in the tropics. Coral reefs form in a belt within 30° north and south of the equator. The corals need a constant supply of warm, nutrient-rich water for their survival. In the tropics, the major ocean currents move from east to west; therefore, the major areas of coral growth are concentrated along the western margins of the oceans. The corals will take hold on almost any type of a hard substrate, so corals rapidly become established on rocky shores or on the hardened lava flows of dormant volcanoes. Once the coral reefs are present, they protect the shoreline from wave erosion.

When a hard substrate is absent, or the turbidity near river mouths is too high for coral growth, other types of organisms inhabit the coastal zone. In protected bays along tropical coasts, mangrove trees extend their roots into the bottom and form a wave-resistant mass. The mangrove roots trap sediment that builds up the coast and extends the shoreline seaward (Figure 1–15). In some tropical areas, Serpulid worms secrete hardened tubes that become intertwined to form reefs.

FIGURE 1–15. Mangrove trees taking root on the beach at Saint Johns in the U.S. Virgin Islands. *Photo by Joan Edwards.*

Mud coasts. Finally, one of the most unusual coastal types is the mud coast. On the coast of Surinam along the northeast margin of South America, a long section of the coast is formed from a slowly moving mass of mud. The mud has a wave form that slowly undulates along the shore. It is impossible to determine where ocean ends and the land begins, because the mud mixes upward into the overlying water. When waves move from the open ocean across the mud coastline, they slowly lose their shape in the mud and die out completely before they reach land. Although it would be an interesting place to study the behavior of waves, researchers have been slow to rush in because of the uninviting environment that, incidentally, is also inhabited by 5-foot mud sharks.

Summary

Each of us has his or her own impression of a coast based on personal experience, but coasts vary widely from one place to the next. Our impressions of the coast are guided by the observations we make at the shore. Studying and sketching coastal features and organisms will enhance our power of observation.

The coastal features that we observe are the product of geologic and biologic processes operating through millions of years. The major structures of the coast are produced by plate tectonics or seafloor spreading. Rigid plates move slowly about on the surface of the earth, driven by forces within the earth. Coasts formed on the leading edges of plates are characterized by rugged shores with long straight cliffs. Coasts on the trailing edges of continents have wide continental slopes and contain abundant barrier islands or reefs.

Many coastal features are affected by changes in sea level. Local sea level changes affect small areas for a short period of time. Regional uplift or downwarp covers a larger area for a longer time. Worldwide or eustatic sea level changes affect the entire ocean for hundreds or thousands of years and are caused by shifting continents or climatic changes such as glaciation.

The edges of the continents are sculptured and molded by rivers, glaciers, and earthquakes. A Ria coast was formed by a drowned river valley due to a rise in sea level. A fjord was carved by a glacier which reaches down to the sea, gouging out a *U*-shaped valley. A karst coast was derived from solution of limestone, and a delta was deposited at the mouth of a river. Volcanic coasts were formed by the eruption of volcanic islands. Secondary coasts were results of wave action or coral reef growth.

Selected Readings

ANIKOUCHINE, W. A. AND R. W. STERNBERG. *The World Ocean—An Introduction to Oceanography* (2nd ed.). Englewood Cliffs, N.J.: Prentice-Hall, Inc., 1981. Concepts of seafloor spreading and plate tectonics are covered in detail.

DAVIES, J. L. *Geographical Variation in Coastal Development* (2nd ed.). London and New York: Longman, 1980. A clear discussion of the different types of coasts on continental margins.

GROSS, M. G. *Oceanography, A View of the Earth* (3rd ed.). Englewood Cliffs, N.J.: Prentice-Hall, Inc., 1982. A comparison of Atlantic-type (Passive) and Pacific-type (Active) continental margins.

SHEPARD, F. P. *Geological Oceanography: Evolution of Coasts, Continental Margins and the Deep Sea Floor.* New York: Crane, Russak and Company, 1977. A geologist's view of different types of coasts.

SHEPARD, F. P. AND H. R. WANLESS. *Our Changing Coastlines.* New York: McGraw-Hill, Inc., 1977. A geologic and photographic tour of the coasts of North America.

SULLIVAN, W. *Continents in Motion.* New York: McGraw-Hill, 1974. An excellent introduction to plate tectonics and seafloor spreading for the general public.

WEILHAUPT, J. G. *Exploration of the Oceans, an Introduction to Oceanography.* New York: Macmillan, 1979. Explanation of changes in sea level through time.

WILSON, J. T. *Continents Adrift and Continents Aground.* Readings from *Scientific American.* San Francisco: W. H. Freeman and Co., 1976. A collection of original papers on continental drift for the educated layperson.

fair weather sailing

TWO
Coastal weather and climate

Introduction to Coastal Weather

Weather plays an important role in many of man's activities along the coast. Fishermen set their nets and bring in the fish when the weather is good, but they are forced to stay in port when the winds are strong and the seas are high. Beaches are crowded with swimmers and sun worshippers on bright summer days, but fog or clouds drive them indoors. Sailors hoist the main and jib and set the spinnaker when winds are favorable, but when the wind dies, they often have to use their motor or paddle to shore. Since so many of our coastal activities depend on the weather, we are often more aware of the weather along the coast.

Storms along the coast can range from a minor inconvenience to a major tragedy. A summer thunderstorm will send the bathers scurrying for protection. A squall can move rapidly across the water, upsetting sailboats and fishing boats in its path. Hurricanes that are spawned in the tropics during the late summer and early fall can play havoc along the coast. High winds destroy structures that are often flooded by the storm surge. The highest winds and greatest rainfalls recorded in the United States are associated with hurricanes. More than 6000 lives were lost in the Galveston hurricane that hit the Texas coast in 1900.

The ocean also has a moderating influence on the weather along the coast, which makes the coast an attractive place to live for a good part of the year. During the summer, the land heats up more rapidly than the sea. Cool sea breezes blow onshore during the afternoon and lower the temperature about 5 to 8 degrees Centigrade (10 to 15 degrees Fahrenheit). During the winter, the water is generally warmer than the air. The strip of land along the coast has a later autumn, a less severe winter, and an earlier spring than the area 10 to 15 kilometers inland. The growing season along the coast is extended several weeks because the freezes in the late spring and frost in the fall are pushed back toward winter.

Predicting capabilities of the National Weather Service are constantly improving with the use of satellites and computers, but it is still possible for individuals to make local weather predictions based on cloud types and changes in wind direction. Sometimes, it is possible to make more accurate predictions along a coast when you are familiar with the local wind patterns and storm tracks.

Climate is defined as the average weather conditions for an area over at least 30 years. Coastal climates vary with latitude, position on the continent, and alignment of adjacent mountain ranges. Coastal climate is often the determining factor that makes some coasts unusually attractive places to visit, and others definite places to avoid.

Elements of Coastal Weather

The major elements of weather we encounter along the coast include air temperature, clouds, fog, wind, and precipitation. The ocean has a moderating influence on the air temperature that is felt several kilometers inland. Clouds, fog, and rain are produced from warm, moist oceanic air when it encounters the land.

Winds along the coast are generated by the major weather fronts, but also by the difference in air temperature over land and sea. During the afternoon, hot air over the land rises, drawing cooling sea breezes from the ocean. At night, land cools off more rapidly and land breezes blow from the land toward the sea.

Air temperature

Air temperature along the coast is strongly influenced by the heat-retaining capacity of land and water. The rock and soil on the shore absorb the visible rays of the sun, heating up rapidly and reradiating heat into the atmosphere. During the day, air over the land is heated rapidly, but it also cools off quickly at night. The water in the oceans is heated more slowly and retains the heat for a longer period of time.

Under similar conditions, the air temperature rises faster and to a higher degree over land than over water. Along the coast and for about 15 kilometers inland, air temperature changes from day to night and from summer to winter are less than they are over continental regions. The temperature changes from one day to the next are also smaller in coastal regions than in the interior of a continent. The large heat storage capacity of water and circulation in the upper layer of the ocean result in a time lag in reaching the maximum temperature of the water, which is transferred to the overlying air.

Large oceanic currents also affect the air temperatures along the coast. For example, the Gulf Stream current carries warm water northeast across the Atlantic from the Caribbean to western Europe. Subtropical plants survive at 52 degrees

Table 2–1. MAXIMUM AND MINIMUM TEMPERATURES IN DEGREES C FOR BOSTON, LONDON, AND GOOSE BAY.

		JAN.	APR.	JULY	OCT.	AVG.
Boston, Mass.	Max.	2.2	12.2	26.7	16.7	14.4
42°22′N	Min.	−6.7	3.3	16.7	7.8	5.6
London, England	Max.	6.7	13.3	22.8	14.4	14.4
51°29′N	Min.	1.7	4.4	12.8	6.7	6.1
Goose Bay, Canada	Max.	−13.3	2.8	21.7	7.2	4.4
53°21′N	Min.	−22.2	−7.2	11.1	−0.6	−5.0

north latitude along the west coast of Ireland. Boston, Massachusetts, and London, England, have very similar average temperatures, but London is almost 9 degrees latitude north of Boston (Table 2–1). However, winter temperatures are much colder and summer temperatures much hotter in Boston than in London. Goose Bay, Labrador, which is about 2 degrees latitude north of London has winter temperatures that are over 20 degrees Centigrade colder than London, but summer temperatures are almost equivalent (Table 2–1).

Clouds   Clouds, along with changes in wind direction, provide a useful method for short-term weather predictions along the coast. Clouds are recognized by their general appearance and their height. The three basic cloud forms are cirrus, cumulus, and stratus (Figure 2–1).

High clouds.   Cirrus clouds are formed from ice crystals above 6000 meters in altitude. Ice particles in cirrus clouds start to fall as soon as they are crystallized. Upper-level winds stretch

FIGURE 2–1. Classification of clouds according to height and form.
After Ward's Natural Science Establishment, Inc., Rochester, N.Y.

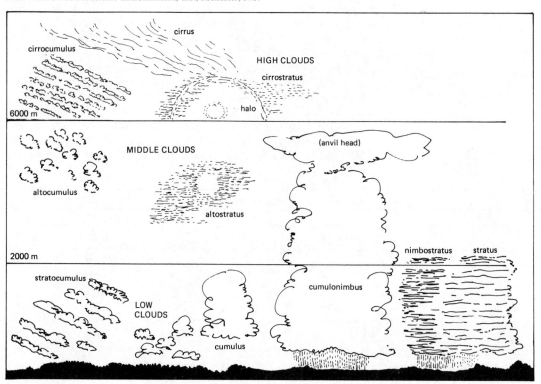

FIGURE 2–2. Ice crystals above 6000 meters form cirrus clouds or mare tails.
Photo by W. T. Fox.

the falling ice crystals out into long streamers. As ice crystals descend to lower altitudes, winds blowing in different directions form a hook or "mares tail" in the cirrus clouds (Figure 2–2).

Cumulus clouds have globular or puffy appearance. The cirrocumulus clouds form ripples, waves, or globular masses from ice crystals above 6000 meters. The cirrocumulus clouds indicate an instability in the upper air without high-speed winds. The cirrocumulus clouds may form a "mackerel sky," which sailors take as a foreboding of bad weather (Figure 2–1).

Stratus clouds are sheets or layers that cover most or all of the sky. The cirrostratus clouds form a thin sheet of ice crystal clouds that give the sky a milky look and form a halo around the sun or moon (Figure 2–1). The refraction of the sun's rays by the ice crystals produce a halo around the sun at 22 degrees angular distance. The halo is a bright ring of light with orange-red on the inside. The different varieties of cirrus clouds usually indicate the arrival of a warm front in 12 to 18 hours.

Middle clouds. The middle clouds at altitudes between 2000 and 6000 meters are a combination of ice and water drop-

33

FIGURE 2–3.
Middle clouds between 2000 and
6000 meters formed by ice and
water are altocumulus clouds,
which look like a flock of sheep.
Courtesy of NOAA.

let clouds. The water droplets in a cloud are about 10 to 50 microns (0.01 to 0.05 millimeters) in diameter. For comparison, a human hair is about 75 microns in diameter.

Altocumulus clouds are white to gray globular masses, like balls of cotton with patches of blue sky between the clouds. The altocumulus clouds are often called "sheepback" clouds (Figure 2–3). Altostratus clouds appear as a thin veil of gray clouds that may produce a very light rain or drizzle. When altostratus clouds are thin, the sun or moon may be visible as a bright spot through the cloud, but no halos are produced. Altostratus clouds give sunlight a filtered appearance like ground glass and are a good indication that rain or snow will arrive in 6 to 12 hours.

Low clouds. The low level clouds extending from the earth's surface up to about 2000 meters include stratocumulus, stratus, and nimbostratus clouds. Stratocumulus clouds are soft and gray with globular patches or rolls (Figure 2–1). Crests of the clouds are often aligned in the direction of the prevailing wind. Stratus clouds form a uniform low layer that resembles fog, but are lifted off the ground. The lower stratus cloud will often produce a light rain or drizzle. When rain becomes a steady downpour, it becomes a nimbostratus cloud (Figure 2–4). The prefix *nimbo-* means rain, so when a stratus cloud starts to produce a steady rain, it is called a nimbostratus cloud.

Vertical clouds. Some clouds have their bases at a low altitude, but their tops extend to the middle or upper range. These clouds are referred to as clouds of vertical development and are associated with unstable air. The bright, cotton-puffy clouds that form on a clear day are called cumulus clouds and indicate fair weather. However, the dark, ominous clouds forming towering thunderheads are cumulonimbus clouds, which can bring heavy rainfall, hail, lightning, and thunder.

FIGURE 2–4.
Nimbostratus clouds below 2000
meters.
Courtesy of NOAA.

The cumulus or fair weather clouds that seem to appear out of nowhere on a pleasant summer afternoon are the product of small, local convection cells or thermals (Figure 2–5). The base of the cloud is usually flat, indicating the condensation level where the rising water vapor in the thermal condenses into fine water droplets. The high concentration of thin cloud droplets reflects the sun's light and gives the clouds a bright white appearance. It seldom rains from cumulus clouds because the cloud droplets settle slowly and evaporate rapidly. As the air in the clouds is carried farther aloft by updrafts in the thermals, the tops of the clouds have a rapidly changing, bulging appearance. The average life span of cumulus clouds on a summer day is about 15 minutes.

Where the moisture content of the air and the afternoon temperature are high, small cumulus clouds build up into large

FIGURE 2–5.
Cumulus clouds formed by strong
updrafts.
Courtesy of NOAA.

FIGURE 2–6.
Anvil top cumulonimbus clouds
during a thunderstorm on the
Florida coast.
Photo by W. T. Fox.

thunderheads or cumulonimbus clouds. When the top of the cumulonimbus cloud reaches above 6000 meters, an anvil head forms as the top of the cloud turns to ice, which is spread out by upper winds (Figure 2–6). The top of a cumulonimbus cloud may reach an altitude of 3000 to 8000 meters (2 to 5 miles). The cumulonimbus clouds are often accompanied by heavy rainfall, lightning, thunder, and tornadoes. Although isolated afternoon thunderstorms often develop in the middle of a maritime tropical air mass, the most severe thunderstorms are found along a cold front where warm moist air is forced aloft by the rapidly advancing wedge of cold air.

Fog. Fog is a common occurrence along many coasts. Fog is simply a cloud with its base at or near the ground. When moist air is saturated with water vapor, and the temperature is dropped, the water vapor condenses and a fog is formed. Warm air blowing over the ocean often contains enough water vapor to form a fog, and the only additional condition necessary is to cool the air.

Most coastal fogs are produced by the upwelling of deep, cold ocean water along the shore. When the wind blows along the shore, the Coriolis effect forces the water to the right in the Northern Hemisphere and to the left in the Southern Hemisphere. An explanation of the Coriolis effect will be given in the following section on wind. If the wind is blowing from the north along the west coast of a continent in the Northern Hemisphere, coastal upwelling will produce a layer of cold water 10 to 15 kilometers wide along the coast. When moisture-laden wind blows over cold water, an advection fog develops that moves onshore with the local sea breeze.

Cape Disappointment on the coast of Washington is one of the foggiest locations in the United States, and perhaps the world. Upwelling during summer and fall produces an average of 2552 hours, or 106.3 days, of fog per year. Fogs are also common on the New England coast, where the cold Labrador current meets a warm, moist air from the Gulf of Mexico.

Wind

Wind has an important influence along the coast, where it affects the local waves and weather conditions. Winds are described by the compass direction from which they are coming, in contrast to currents, which are given in the direction they are heading. The wind speed is given in miles per hour or knots (nautical miles per hour) or described by Beaufort Force. The relationships between the Beaufort Force and the wind speed in knots and miles per hour are given in Table 2–2. Criteria are also given for determining the Beaufort Force on land and at sea in Tables 2–3 and 2–4.

Along the coast, it is better to use the land criteria, because the sea criteria are valid only for deep-water waves. Breaking waves can form along the coast when the wind speed is practically zero, due to the shoaling effects. The effect of wind on the sea depends on the wind speed, fetch, and distance it has blown over the water. Currents in the open ocean also add to or subtract from the wind effects. When the wind is blowing against a strong current, such as the Gulf Stream, the speed of the current is added to the effective wind speed, producing a higher Beaufort Force. When the wind is blowing in the same direction as the current, a lower apparent wind velocity would be given by the Beaufort Force.

Table 2–2.   WIND VELOCITY TERMINOLOGY.

BEAUFORT FORCE	WIND (MPH)	SPEED (KNOTS)	NATIONAL WEATHER SERVICE TERMINOLOGY
0	1	1	calm
1	1–3	1–3	light air
2	4–7	4–6	light breeze
3	8–12	7–10	gentle breeze
4	13–18	11–16	moderate breeze
5	19–24	17–21	fresh breeze
6	25–31	22–27	strong breeze
7	32–38	28–33	near gale
8	39–46	34–40	gale
9	47–54	41–47	strong gale
10	55–63	48–55	storm
11	64–72	56–63	violent storm
12–17	73–136	64–118	hurricane

Table 2–3. DETERMINING BEAUFORT WIND FORCE ON LAND.

BEAUFORT FORCE	DESCRIPTION
0	Calm, smoke rises vertically
1	Smoke drifts, but weather vanes don't change
2	Wind felt on face; leaves rustle; weather vane moves
3	Leaves in constant motion; small flags extended
4	Raises dust and loose paper; small branches move
5	Small trees in leaf begin to sway; crested waves on inland waters
6	Large branches in motion; telephone wires whistle; umbrellas used with difficulty
7	Whole trees in motion; resistance in walking against the wind
8	Breaks twigs off trees; generally impeded progress
9	Slight structural damage occurs (chimney parts and slate removed)
10	Trees uprooted; considerable structural damage; seldom experienced inland
11	Widespread wind damage; very rarely experienced inland
12	Maximum wind damage

Table 2–4. DETERMINING BEAUFORT WIND FORCE AT SEA.

BEAUFORT FORCE	DESCRIPTION
0	Sea like mirror
1	Small ripples; cats' paws on surface
2	Small wavelets; crests smooth with glossy appearance
3	Large wavelets with crests beginning to break; scattered whitecaps
4	Small waves growing larger; numerous whitecaps
5	Moderate waves with greater length; many whitecaps with some spray
6	Larger waves; whitecaps more numerous; more spray
7	Seas heap up and foam blown in streaks from waves
8	Fairly high waves of greater length; well-marked streaks of foam
9	High waves with sea beginning to roll; dense streaks of foam; spray may cut visibility
10	Very high waves with overhanging crests; sea is white with foam; heavy rolling and reduced visibility
11	Waves exceptionally high; sea covered with foam; visibility further reduced
12	Sea completely covered with spray; air filled with foam; greatly reduced visibility

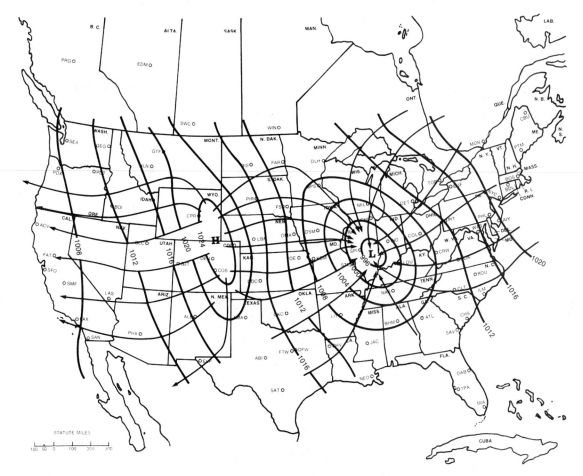

FIGURE 2–7. Pressure-gradient lines move out of the high and into the low.

Pressure-gradient force. Winds on the earth's surface are due to differences in barometric pressure. On a nonrotating earth, wind would follow the pressure-gradient force from areas of high pressure to areas of low pressure. The lines of equal barometric pressure, or isobars, form approximate concentric circles around a low- or high-pressure system (Figure 2–7). Where isobars are close together, the pressure-gradient force is greatest.

Coriolis effect. While the earth is rotating about its axis, winds generated by the pressure gradient force are deflected to the right in the Northern Hemisphere and to the left in the Southern Hemisphere. This deflection is a result of the earth's rotation and is called the Coriolis effect.

The Coriolis effect was first described by Gaspard G. Coriolis, a French engineer and mathematician. In 1824, Coriolis was studying the effect that the rotation of a body had on the ordinary laws of motion. He found that if a body is rotated in a counterclockwise direction, an inertial force to the right must be added to the ordinary forces. Similarly, if the body is rotated in

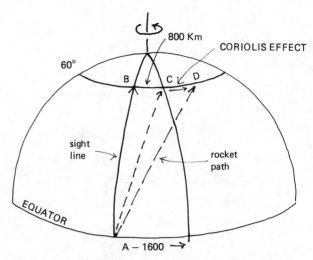

FIGURE 2–8. A rocket launched from A toward B (due north).
Due to earth's rotation, the rocket has an initial speed of 1600
km/hour to east; therefore, it lands at point D, 1600 km east
of B. At the same time, B moves 800 km east to C. Therefore,
the rocket lands at D, 800 km east of the target. It appears
that the rocket has veered to the right 800 km, which is the
Coriolis effect.
Drawing by C. W. L.

a clockwise direction, an inertial force to the left must be added.
When you look down on the earth's North Pole, the earth rotates
in a counterclockwise direction, so a force must be added to the
right for any motion in the Northern Hemisphere.

The Coriolis effect is often demonstrated by considering a
rocket launched from the equator toward the North Pole (Figure
2–8). As the earth rotates about its axis for 1 hour, the launch-
ing platform on the equator moves to the east 1600 kilometers.
The target located at 60° north latitude moves 800 kilometers
eastward at the same time. The lines of longitude converge toward the poles, so that the distance traversed by any point on
the earth's surface due to the earth's rotation is less toward the
poles. When a rocket is launched due north, the rotation of the
earth also gives it an eastward component of 1600 kilometers
per hour. When the rocket lands at 60° north latitude one hour
later, the target has moved eastward only 800 kilometers and
the rocket lands an additional 800 kilometers to the east. As
viewed from the earth, the path of the rocket has veered to the
right of the target. By using similar logic, it is possible to prove
that the rocket would veer to the right if it was fired from 60°
north due south toward the equator. Similarly, if a rocket is
fired to the east or west in the Northern Hemisphere, it will always veer to the right.

As Coriolis described the effect, it involves the frame of
reference from which you are viewing the motion. When a
rocket launched from earth is viewed from space, it travels

along a straight path without deflection. However, when the path of the rocket is viewed from the surface of the earth, it would veer to the right in the Northern Hemisphere and to the left in the Southern Hemisphere. In considering winds on the earth's surface and currents in the oceans, a deflection to the right must be added for the Coriolis effect in the Northern Hemisphere.

Geostrophic winds. Winds in the upper atmosphere from 5 to 10 kilometers above the earth's surface are known as geostropic winds. Air moving from the center of a high-pressure area into the center of a low due to the pressure gradient force is deflected to the right by the Coriolis effect until it flows parallel to the isobars. In the Northern Hemisphere, geostrophic winds blow in a counterclockwise direction around lows and clockwise around highs (Figure 2–9).

Surface winds. The angle and speed of surface winds are affected by friction over land and sea. At 40 degrees latitude, angles between the theoretical geostrophic wind direction and the surface wind direction are about 42 degrees over land and 16 degrees over sea. The wind angle is always bent in toward the center of a low and out from the center of a high-pressure system (Figure 2–10). The land has a greater effect in slowing down the surface wind than the sea, due to the greater irregularities and friction over the land. At 40° latitude, the surface wind speed is about .38 times the theoretical geostrophic wind speed over land, and about .63 times geostrophic wind speed over open water.

The counterclockwise wind pattern around the center of a low in the Northern Hemisphere is referred to as cyclonic circulation. The clockwise circulation around a high pressure is considered anticyclonic. All major storms are cyclonic storms circulating around low-pressure systems. If you face into the wind blowing around a cyclonic storm and extend your right

FIGURE 2–9. Geostrophic winds in the upper atmosphere are deflected to the right by the Coriolis force until the Coriolis force just balances the pressure-gradient force. Above 600 meters, the geostrophic winds flow parallel to the isobars.

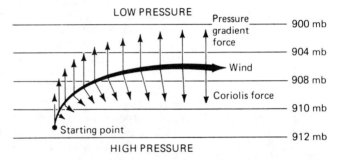

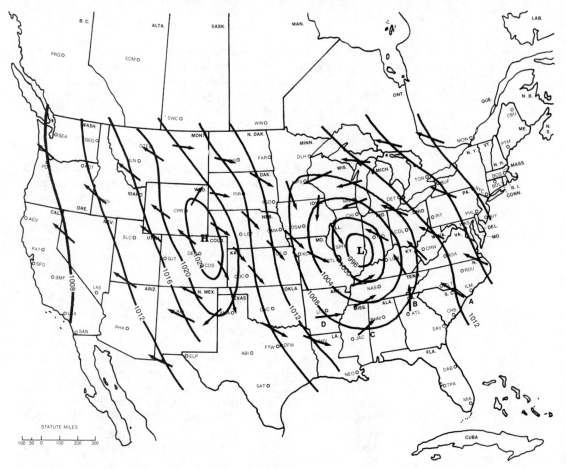

FIGURE 2–10. Surface winds move in a clockwise motion
and out from a high, and move in a counterclockwise motion
and into a low.

arm, you are pointing in the general direction of the low-pressure center. In the Southern Hemisphere, the Coriolis effect causes a deflection to the left. Therefore, a cyclonic storm circulates in a clockwise direction, south of the equator.

The surface wind is quite variable in speed and direction due to eddies near the ground. When an eddy dips down to the earth's surface, the wind speed will pick up and change direction, dying off again as the gust rises. The gustiness of the wind should not be confused with a steady change in wind direction, which can be useful in predicting local weather changes.

The wind may shift in either a clockwise or counterclockwise direction. When a low-pressure center passes to the north, the wind shifts in a clockwise direction from east through south to west and northwest. The shift in compass directions indicates a veering wind (Figure 2–11A). However, when a high-pressure system passes to the north of the observer, the wind direction shifts in a counterclockwise fashion from west through south to east, producing a backing wind (Figure 2–11B). The old weather

42

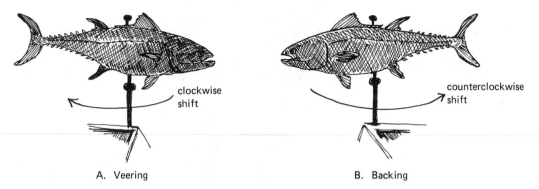

A. Veering   B. Backing

FIGURE 2–11. A veering wind (A) shifts in a clockwise direction (east through south to west), and a backing wind (B)
shifts counterclockwise (from west through south to east).
Drawing by C. W. L.

proverb, "a veering wind means weather fair, a backing wind
foul weather near," is often, but not necessarily, true. A veering
wind indicates a passing low with clearing weather on its way.

Land and sea breezes.   One of the attractive features of
living along the coast is the land and sea breezes, which modify
the daily temperatures. During the afternoon in the summer
months, the land heats up more rapidly than the sea. As air over
the land heats, it expands and rises, forming a high-pressure
zone aloft (Figure 2–12A). The air over the sea at a higher level
is at a slightly lower pressure, and air flows from land to sea.

FIGURE 2–12. A sea breeze (A) is formed in the afternoon by
expanding warm air that rises over the land and draws in the
cool air from the sea; and a land breeze (B) blows seaward at
night when the land cools more rapidly than the sea.
Drawing by C. W. L.

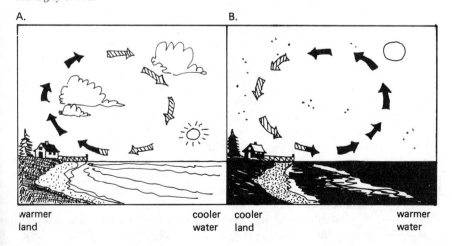

A.   B.

warmer   cooler   cooler   warmer
land   water   land   water

With the addition of air aloft over the sea, a high-pressure area forms at the surface. Therefore, the air at the sea surface follows the pressure gradient and flows toward the land. The air circulation along the coast forms a convection cell with the upper air moving out to sea and the surface air blowing toward the land. Since the sea is generally cooler than the land during the day, the sea breezes bring relief from summer heat. The temperature along the coast is usually 5 to 10 degrees Centigrade (9 to 18 degrees Fahrenheit) cooler than temperatures 5 to 10 kilometers inland, which are unaffected by the sea breeze.

At night, air over the land cools more rapidly than the sea. The sea has a higher specific heat and therefore retains the heat into the night. As the warmer air rises off the sea, the land breeze blows toward the sea with a return flow toward the land at the upper level (Figure 2–12B). The difference in temperature between land and sea is greatest during the day; therefore, the sea breeze is usually considerably stronger than the land breeze. The sea breeze during the afternoon blows landward at 10 to 15 knots, while the land breeze at night is 3 to 5 knots. As the temperature drops over the land at night, the warmer air along the coast moderates the drop in temperature.

Weather Systems

Weather in the mid-latitudes is controlled by large weather systems that move from west to east across the surface of the earth. The weather systems are dominated by large air masses of either cold, dry air or warm, moist air. Frontal systems and storms usually form along the boundary between the large air masses. Weather forecasting is made possible by plotting the movement of air masses and fronts on weather maps.

Air masses

Weather that directly affects the coasts is produced from the interaction of large, relatively stable air masses that form over the continents and oceans. Air masses, which are usually more than 1000 kilometers across and several kilometers thick, are characterized by similar temperature and moisture content at any particular altitude. Because individual air masses may span 10 to 20 degrees of latitude, there are small differences in temperature and moisture within an air mass. However, differences within an air mass are minor compared with the major differences observed between adjacent air masses.

Weather within an air mass is relatively stable, and the unsettled weather or storms congregate along the borders, or "fronts," between air masses. As air masses grow in size, they tend to spread out and intrude on other air masses. The cold, dry polar air masses are more dense and wedge themselves beneath the warm, moist air masses of the tropics. When warm air is lifted into the atmosphere, moisture condenses, forming clouds, and eventually rain or snow.

Stable areas in which the major air masses originate are called source regions (Figure 2–13). The source regions for large

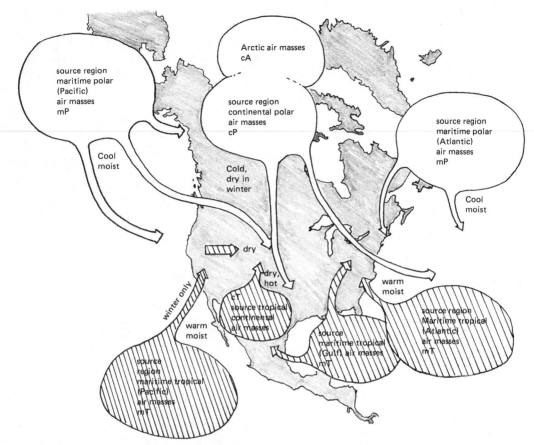

FIGURE 2–13. Air mass source regions of North America.
Courtesy of Ward's Natural Science Establishment, Inc. Rochester, N.Y.

air masses are located in the middle of continents or over oceans. The continental air masses *(c)* have a low water vapor content, producing dry air, and the maritime air masses *(m)* have a high water vapor content, giving humid air. Air masses are also divided by latitude into four categories: polar *(P)*, arctic *(A)*, tropical *(T)*, and equatorial *(E)*. The difference between polar and arctic, and between tropical and equatorial are usually small and are used to indicate the extreme cold or hot air masses. The following air masses are common on the earth.

cA continental arctic
cP continental polar
cT continental tropical
mT maritime tropical
mP maritime polar
mE maritime equatorial

45

The *mA* maritime arctic and *cE* continental equatorial air masses are not listed because they are seldom formed from the distribution of continents and oceans at the correct latitudes.

The principal North American air masses were plotted to show their source regions (Figure 2–13). The continental polar *(cP)* and continental arctic *(cA)* air masses contain cold and dry air, which originates north of 55° latitude. The continental polar air masses form on the snowfields of interior Canada and Alaska. The continental arctic air masses come from the Arctic Basin and the Greenland ice cap. In winter, continental polar and arctic air masses invade the United States from the north, giving us bone-chilling deep freezes in January and February. In January, 1977, large arctic air masses covered eastern North America for several weeks, resulting in freezes on the Great Lakes, Long Island Sound, Delaware Bay, and Chesapeake Bay. The cold air spread as far south as Florida, nipping the orange buds and ruining a large portion of the citrus crop. The continental arctic air masses reach the United States only during the winter, while continental polar air masses are responsible for refreshing cold fronts during the summer.

Maritime polar air masses *(mP)* form at high latitudes over the oceans. The two source regions of maritime polar air are the north Pacific, and the northwest Atlantic from Cape Cod to Labrador (Figure 2–13). During winter, maritime polar air from the north Pacific gives rise to rain and cold, damp air along the Pacific coast. Washington and Oregon have from 70 to 100 inches of rainfall during the winter, with heavy snowfall in the Cascade Mountains.

During the summer, a high-pressure cell, which feeds cooler air from the north along the coast, stabilizes off the west coast of the United States. Almost continuous north and northwest winds along the Pacific coast generate a southward moving surface current and coastal upwelling. When the cold water brought up by upwelling meets the warm summer air, coastal fog develops. The California, Oregon, and Washington coasts, which are bathed by rain during the winter, are shrouded by fog during the summer. The best time to visit the Washington or Oregon coasts is in late spring or early fall between the rain and fog seasons.

Maritime tropical *(mT)* air masses originate in the warm waters off the Gulf of Mexico and the adjacent western Atlantic Ocean and Caribbean Sea (Figure 2–13). During the winter, continental arctic and polar air masses dominate the central and eastern states, and maritime tropical air masses are forced to the south. During the summer, the sun moves north of the equator, increasing the size and influence of the maritime tropical air masses. The moisture from the Rocky Mountains east to the Atlantic coast is provided by the warm, moist air from the Gulf of Mexico.

Weather fronts   The boundaries between warm and cold air masses are known as fronts. The concept of fronts is based on the work of a group of Norwegian meteorologists during World War I, who compared the air masses to large armies and the boundaries to fronts that marked the zone of conflict.

As a continental polar mass moves toward the equator from the pole, it advances as a tongue of dense, cold air protruding beneath the warm, moist tropical air mass. A series of waves are formed along the polar front, which is the boundary between the continental polar and maritime tropical air masses. The waves slowly move from west to east, giving rise to the steady progression of fronts that are experienced in the summer months.

Cold fronts. A cold front is formed when the leading edge of a cold air mass forces its way beneath a warm moist air mass (Figure 2–13). The average cold front is relatively steep, with a slope of about 1 to 100. Violent weather, including thunderstorms, squalls, and tornadoes, is often associated with cold fronts. The arrival of a cold front is often preceded by nimbocumulus clouds, towering clouds that can be seen in the northwest. When there is a strong contrast between warm moist air and a rapidly moving cold air mass, a fast moving squall line forms about 250 to 300 kilometers ahead of the cold front. Tornadoes and water spouts are often generated in the severe thunderstorms accompanying a squall line. The passage of a cold front is marked by rapid drop in temperature, often 20 to 40 degrees Fahrenheit, in less than an hour. After the cold front has passed, the wind shifts to the northwest and the skies clear as the cool, dry air enters the region.

Warm fronts. A warm front is located where the warm, moist maritime tropical air mass takes over the space vacated by the cold, dry polar air mass (Figure 2–14). As the wave of cold air moves to the east due to the rotation of the earth, warm air moves up from the Gulf of Mexico to replace the departing cold air. The trailing edge of cold air forms a thin wedge with a slope of about 1 to 200. The warm, moist air moves up over the thin edge of cold air, rising 1000 meters vertically for every 200 kilometers on the ground. Therefore, the clouds preceding the warm front are seen about 1000 to 1200 kilometers ahead of the actual front.

As a warm front moves across a region, the usual sequence of clouds would be cirrus, altostratus, stratus, and nimbostratus, as the base of the clouds drops in altitude and precipitation changes from light drizzle to steady rain. The rain usually starts about 250 to 300 kilometers in advance of the warm front and continues until after the front has passed.

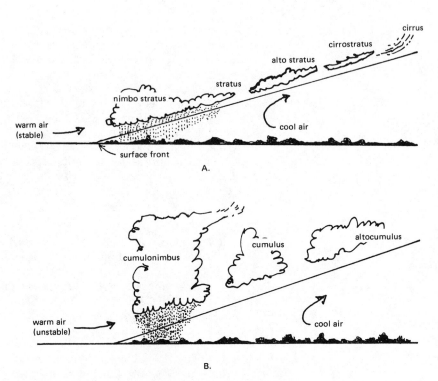

FIGURE 2–14. When a warm air mass flows over a cold air mass, clouds form as the warm air is lifted aloft.
Drawing by C. W. L.

Occluded fronts. When a cold front overtakes a warm front, an occluded front is produced. The cold front generally advances more rapidly than a warm front and eventually shoves beneath the tail of the warm front. As the active cold front advances beneath the retreating warm air mass, the warm air is trapped and carried aloft, producing heavy rain. When the air behind the cold front is colder than the retreating warm air mass, which is usually the case east of the Rockies, a cold front occlusion takes place in which the advancing wedge moves under the retreating warm front (Figure 2–15). Along the Pacific coast, the air behind the advancing cold front is often warmer than the retreating cold air mass, which may have come down from Canada, and a warm front occlusion is produced. In a warm front occlusion, the advancing cold front overrides the denser retreating air mass.

Stationary fronts. When a frontal system becomes stalled and stops moving forward, it becomes a stationary front. The winds north and south of the front are blowing in parallel but opposite directions. When a stationary front starts to move once again, it will become a cold front if it bulges to the south or a warm front if it pushes north.

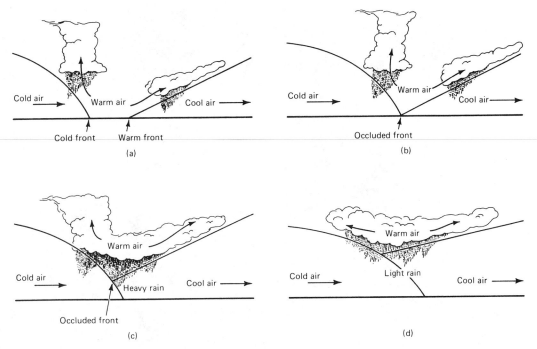

FIGURE 2–15. The formation and eventual dissipation of an occluded front.

Wave cyclones   The Norwegian model of a wave cyclone, based on surface measurements made during World War I, is useful for understanding daily weather patterns and making local weather predictions. As our knowledge of meteorology has advanced with weather balloons and satellite observations, it is now possible to make more accurate weather predictions from high-altitude pressure maps. However, for day-to-day forecasts, the wave cyclone model makes it possible to predict wind and weather changes for 24 to 36 hours in advance of a storm, a useful procedure along a coast.

A wave cyclone develops from a stationary front through a low-pressure center with dangling warm and cold fronts to an occluded front (Figure 2–16). Along the stationary front, the wind in the warm air to the south of the front blows to the east, while the wind north of the front in the cold air mass is blowing to the west. As cold air starts to move south, a low forms in the stationary front with a cold front to the west and a warm front to the east. Wind begins to blow around the low in a counterclockwise direction. As the cyclone develops, the leading warm front advances to the northeast and the trailing cold front moves to the southeast. As the circular wind pattern strengthens, the low-pressure cell deepens at the bend in the frontal line. Upper air winds are blowing away from the low, causing a drop in barometric pressure. As the cyclone continues to develop, the

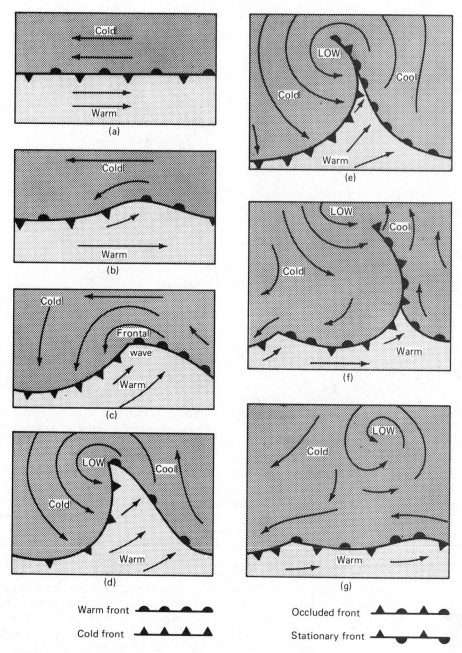

FIGURE 2–16. Stages in the life cycle of a middle-latitude cyclone as proposed by J. Bjerknes.

cold front advances more rapidly than the warm front, forcing an acute angle in the frontal line at the center of the low (Figure 2–16). When the cold front catches up with the warm front, an occluded front pushes the warm air aloft (Figure 2–16). At this stage, the cyclone is the most dangerous, with strong winds and heavy rains. When the occluded front continues its advance, the low dissipates as warm, moist air is carried aloft. In the last stage, the low-pressure center becomes a weak counterclockwise circulation in the upper air, with a new stationary front forming to the south.

Often a series of waves will develop along a frontal line, with successive cyclones forming at each wave position. Along the front, the cyclones progress from a stationary front through an occluded front, each cyclone changing its character as it moves to the east. By watching the progression of fronts on a series of weather maps, it is often possible to forecast where an occlusion and eventual dissipation of a cyclone will take place.

The position of the jet stream along the polar front determines the path that cyclones will follow from west to east. When the jet stream follows a relatively straight path from west to east across the upper part of the United States, the cyclones move along an east-west track, producing expected amounts of rainfall across the upper tier of states. However, when the jet stream follows a curved path and dips to the south across the Rocky Mountains, the storm track moves to the south, leaving the northeast with extremely cold conditions in winter and drought in summer. When the jet stream shifts back to its more normal position, the drought conditions are alleviated, but when it persists for several months, severe drought and crop damage may result. This occurred during the winters of 1976–77 and 1977–78, when the severe drought conditions prevailed.

Coastal Storms

Most of the high waves and beach erosion occur during coastal storms. Thunderstorms and squalls are frequent in the mid-latitudes and are extra tropical cyclones. Tropical cyclones that develop into hurricanes are formed in the tropics and move into higher latitudes. High winds and heavy rains that accompany hurricanes cause considerable damage along the coast. Coastal flooding is also caused by storm surges when water is piled up along the shore by a hurricane.

Thunderstorms and squalls

Thunderstorms and squalls are some of the most violent storms along the coast. Strong, gusty winds; hail; lightning; and rapid buildup of seas are especially dangerous for small craft. Thunderstorms are the result of large convection cells that rapidly build up and dissipate their energy over a small area. Towering cumulonimbus clouds rising up to 8 kilometers (5 miles) in the sky are evidence of large and potentially dangerous thunderstorms.

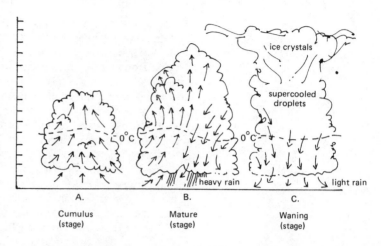

FIGURE 2–17.
Stages in the development of a
thunderstorm. (A) Cumulus stage
with strong updrafts; (B) mature
stage with heavy rain, updrafts,
and downdrafts; and (C)
dissipative stage with light rain,
weak downdrafts, and an
anvil top.
Drawing by C. W. L.

A.
Cumulus
(stage)

B.
Mature
(stage)

C.
Waning
(stage)

Thunderstorms generally pass through three stages—
cumulus, mature, and dissipation—within a few hours. The cumulus stage of thunderstorm development starts with a small
convection cell (Figure 2–17A). Mixing of moist air in small cumulus clouds and cool, dry air aloft results in evaporation at the
top and sides of the cloud, and most initial cumulus clouds disappear in 10 to 15 minutes. Late in the afternoon, when heating
becomes more intense, a few small cumulus clouds entrain
warm, moist air from the sides and build up into large thunderheads. The process of entrainment allows clouds to feed in
more warm, moist air, which is essential for the formation of a
thunderhead. The cumulus stage generally lasts about 15 to 20
minutes while the cloud is dominated by updrafts. The updrafts
within the cumulus stage may reach 60 kilometers per hour in
the upper portion of the cloud.

The mature stage begins when rain starts to fall out of the
cloud (Figure 2–17B). Once the upper portion of the cloud
passes the freezing level, ice crystals in the cloud are the nuclei
for raindrops. As the cloud mushrooms upward, cool, dry air is
incorporated into the cloud by entrainment. The dry air causes
some of the falling rain to evaporate, which further cools the
air. The cool air forms a rapid downdraft within the cloud,
which carries the precipitation with it. At the ground surface,
cool air spreads out before rain reaches the ground. In a thunderstorm, you will usually notice a cool downdraft shortly before rain arrives. In the mature stage, the downdrafts are usually adjacent to updrafts, which are still sucking warm, moist
air into the cloud. Hailstones, which start in the upper part of
the cloud, may make several roundtrips in downdrafts and updrafts, adding a new layer of ice with each circuit.

When the cloud reaches the base of the stratosphere, updrafts spread out to each side, producing the characteristic anvil
top that indicates the waning phase of the thunderstorm (Figure
2–17C). Ice-laden cirrus clouds in the anvil head are spread out

by high-speed upper-level winds. In the dissipating stage, down-drafts encourage entrainment of cool, dry air, which in turn increases the strength of the downdraft. When the supply of moisture is cut off by dominance of downdrafts, heavy rain turns to drizzle and the thunderstorm dies out.

The life cycle of a single thunderhead from cumulus through mature and dissipating stages usually lasts about an hour or less. Lightning, thunder, and heavy rain are concentrated in the mature stage, which lasts about 20 minutes to half an hour. However, as the thunderstorm activity of a convection cell grinds to a halt, it is replaced by other cells that take up the warm, moist air. On a hot summer evening, it is often possible to see lightning and rain from several mature cells, while some are just starting their growth, and still others are dying away.

Frontal thunderstorms are formed when a steep cold front forces its leading edge beneath a moist, warm air mass (Figure 2–18). The squall lines move at about 40 kilometers per hour where the convection cell formed, frontal thunderstorms move in a line across the countryside and are easier to predict. Tornadoes and squalls are often associated with frontal thunderstorms and are the cause of considerable damage along the coast.

A squall line often precedes a cold front by 250 to 300 kilometers (about 200 miles). The downdraft from the thunderstorms along the cold front produces a pseudo-cold wave, which acts as a bow wave, forming the squall line. The squall line has a distinctive dark gray cigar-shaped cloud that appears to roll across the sky from one end of the horizon to the other (Figure 2–18). The squall lines move at about 40 kilometers per hour (21 knots). Winds in the squall can range from 60 to 100 kilometers per hour (32 to 53 knots), but they are short-lived. High winds and heavy rains of a squall are usually over within 10 to 15 minutes, but they can raise havoc with small boats. On the Great Lakes, where squalls are a common summer occurrence, a squall line stretching from northeast to southwest is an awesome sight. As the squall moves across the lake, it rapidly

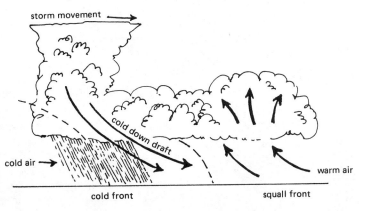

FIGURE 2–18.
A frontal thunderstorm with a
squall line formed along a
cold front.
Drawing by C. W. L.

FIGURE 2–19.
Paths of tropical cyclones and
hurricanes in different parts of
the world.

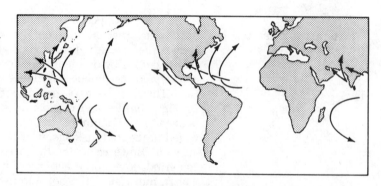

builds up the waves. Within minutes after the squall arrives, the waves increase to about 1 meter. Since the wind in the squall is short-lived, the waves also die down almost as rapidly as they build up.

Florida has a higher concentration of thunderstorms and lightning than anywhere else in the United States. Tampa, Florida, has an average of about ninety thunderstorms per year. The converging warm air masses from both sides of the Florida peninsula give rise to the large numbers of thunderstorms.

Tropical cyclones   Tropical cyclones are storms that develop over the open ocean with energy supplied by warm, moist air and motion provided by the Coriolis effect. For a tropical cyclone to form, surface water temperature must be above 26.5 degrees Centigrade (80 degrees Fahrenheit), and latitude between 5 degrees and 20 degrees north or south of the equator. Along the equator, the horizontal component of centrifugal force that produces the Coriolis effect is nonexistent. Moving north or south about 5 degrees, the Coriolis force is sufficient to start the rotary motion that drives a tropical cyclone. In late summer and early fall when the meteorological equator or intertropical convergence zone (I.T.C.Z.) is between 5 degrees and 15 degrees, the ocean is hot enough to supply moist, warm air for the formation of tropical cyclones.

Tropical cyclones that are spawned each summer must pass through four stages to become full-fledged hurricanes. In the Northern Hemisphere, easterly waves within the northeast trade winds move from east to west with an average speed of 24 kilometers per hour (13 knots). The easterly waves that are spaced out about every 15 degrees around the earth are associated with clouds and thunderstorms.

Low-pressure areas along the easterly wave are known as tropical disturbances. Northeast and southeast trade winds impinging on the tropical disturbance produce a shearing action, forming a rotary motion around the low. As thunderstorm activity builds up within the low, but wind speeds are still below 60 kilometers per hour (32.4 knots), it is considered a tropical depression. Rotating winds feed warm, moist air into the depression, increasing the intensity and forming a tropical storm. A tropical storm with a well-developed rotary motion has wind speeds between 60 and 120 kilometers per hour (32.4 to 64.8 knots). At this stage, the tropical storm usually breaks away from the I.T.C.Z. and is carried westward by the northeast trade winds. The rotation of the earth drives the storm away from the equator, giving it a northwest path.

Hurricanes. When the wind in the storm exceeds 120 kilometers per hour (64.8 knots), the storm is called a hurricane. In the Atlantic, hurricanes are spawned off the west coast of Africa and move to the west and northwest through the Caribbean into the Gulf of Mexico or into the North Atlantic (Figure 2–19). As the hurricane moves north of the horse latitudes (about 30° north) it is caught in the westerlies and follows a path to the northeast. In the northwest Pacific, the storms forming east of the Philippines and moving toward the coasts of China and Japan are known as typhoons (Figure 2–20). In the southwest Pacific, storms forming south of the equator and sweeping the coasts of Australia and New Zealand are known as willy-willies. In the Indian Ocean, the storms that move north toward India and south along the East African coast are called cyclones. In the international classification of tropical storms, all tropical storms with a wind speed greater than 120 kilometers per hour (64.8 knots) are lumped together as hurricanes. The name comes from the Spanish word *huracán*, which was derived from the Mayan storm god *Hunraken.*

A hurricane has a regular internal structure that is apparent from the cloud pattern and wind system. Thunderheads radiate out from the center in spiral arms known as rain bands (Figure 2–21). The eye is an area of relative calm in the center of the storm, marked by extremely low pressure and slowly rising warm air. A solid wall of clouds, several thousand meters thick, surrounds the eye and contains the highest winds. A hot convective chimney forms within the wall of clouds around the eye. The rising hot air that is responsible for the high winds in the wall rises through the chimney and does not flow into the eye. Hot air released through the chimney at high altitudes accounts for the sudden drop in pressure near the center of the storm. Winds flowing across the pressure gradient are turned to the right by the Coriolis effect and reach velocities greater than 300 kilometers per hour.

A hurricane can be thought of as a vast heat machine that sucks in water vapor evaporated from the surface of the sea and converts it into heat energy when the water vapor is condensed into moisture in the form of clouds. Each cubic centimeter of water vapor condensed in the clouds releases 580 calories of heat. The hot air rising in the convective clouds within the rain bands and in the convective chimney forms the pressure gradient within the storm (Figure 2–21). As long as water vapor is

FIGURE 2–20. Satellite photographs of three tropical cyclones:
Hurricane Kate, a tropical depression, and Hurricane Lisa in
the Pacific Ocean on September 26, 1976.
Courtesy of NOAA/EDIS/SDSD.

available from the surface of the ocean, the storm will continue
its destructive growth. However, when a hurricane passes over
a continent, the supply of moisture is cut off and the storm
starts to fizzle. It been estimated that the energy within a
typical hurricane is equivalent to the energy released from 400
20-megaton bombs exploding on the same day, or to all the electrical energy used in the United States in six months.

56

FIGURE 2–21. Satellite photograph of Hurricane David on August 29, 1979 in the Gulf of Mexico.
Courtesy of NOAA/EDIS/SDSD.

Hurricane destruction. Hurricanes inflict most of their destruction in the form of high winds, heavy rains, storm surges, and accompanying tornadoes. A small intense hurricane similar to Camille, which roared through the Gulf of Mexico in 1969, packs a high wind and is usually accompanied by large storm surges, but has a limited area of heavy rainfall. A broad shallow hurricane similar to Agnes, which struck the eastern third of the United States in 1972, has fairly low winds and a small storm surge, but dumps large quantities of rain over a broad area. Tornadoes generated along the leading edge of

57

a hurricane are often destructive, but their effects are usually masked by the high winds surrounding the eye of the storm.

Hurricane Camille is a typical example of a small intense hurricane that can result in major damage along a coast. In August, 1969, Camille moved ashore on the coast of Mississippi with wind gusts of 320 kilometers per hour (about 200 knots or 190 miles per hour). The pressure in the eye of the hurricane was 901 millibars (26.61 inches of mercury), making it the third lowest recorded in this country. [The lowest recorded pressure (892 millibars) was in a hurricane that passed over the Florida Keys in 1935. The second lowest pressure in the Atlantic was reached during hurricane Allen when it dropped to 899 millibars (26.55 inches) on August 7, 1980, as it passed through the Yucatan Peninsula enroute to Texas. The lowest pressure ever recorded in any tropical cyclone was hurricane Ida when it passed over Luzon in the Philippine Islands with a pressure of 886.56 millibars (26.18 inches).] All the storms with low pressure in the core have a steep pressure gradient in the wall around the eye, producing extremely high winds with a theoretical maximum of about 320 kilometers per hour.

The storm surge produced by the high winds from hurricane Camille was about 7 meters (20 feet) above normal. Three freighters were carried ashore by the storm surge and stranded when the water receded. The Richelieu Apartments in Pass Christian, Mississippi, overlooked the beach and were the site of a hurricane party. Twenty-five people gathered on the second floor to watch the storm's arrival. Although they were warned of the potential danger by the chief of police, they refused to leave. The apartment house was totally destroyed and only two survivors were found among the rubble. About 200 people lost their lives in Mississippi and damage was estimated at about $.5 billion (Figure 2-22).

After passing through Mississippi, hurricane Camille lost its punch because it was deprived of moisture. The weakened hurricane swept an arc across the southeastern states and once again passed over the Atlantic, regaining its lost moisture. When it returned ashore for the second time, Camille dropped more than 28 inches of rain in 30 hours over much of Virginia. The James River flooded the lowlands, taking an additional 152 lives and causing more than $100 million worth of damage.

Hurricane Agnes originated in the Caribbean region in mid-June, 1972. The storm passed over the Florida Panhandle coastline near Tallahasse on June 19, and swept on a northward path, bringing heavy rain from the Carolinas to New York. The storm was more than 1000 kilometers wide, but the minimum pressure was only 984.1 millibars when it passed over New York City on June 22. The highest average wind speeds were about 65 to 80 kilometers per hour with gusts up to 115 kilometers per hour (69 miles per hour) at Kennedy, South Carolina. The storm surge was about 1 to 2 meters above normal along the Florida coast. The maximum rainfall was 42.6 centimeters (18.21

FIGURE 2–22. A fishing trawler was driven ashore in Biloxi,
Mississippi, during Hurricane Camille in August, 1969.
Courtesy of NOAA.

inches) at Mahantango Creek in Pennsylvania, but more than
93,000 square kilometers received an average of 25.7 centimeters (11 inches) of rain. The death toll due to the storm reached
118 people, and the estimated damage was $3.1 billion, making
it the most expensive natural disaster ever to hit the United
States.

On September 8, 1900, a hurricane that struck Galveston,
Texas, was the worst natural disaster in terms of lives lost in the
United States. The storm surge that accompanied the hurricane
flooded the bay behind Galveston Island and submerged a good
portion of the city. More than 6000 men, women, and children
lost their lives in Galveston, and 2000 more lives were lost elsewhere along the Gulf coast during the storm.

The most devastating storm in recorded history was an Indian Ocean cyclone that crossed the delta of the Ganges-Brah-
maputra River of East Pakistan on November 12, 1970. The low-

lying coast and adjacent islands were flooded by a 4-meter storm surge that cost more than 300,000 lives. The funnel shape of the coast increased the effect of the storm surge, which was produced by the high winds. Heavy rains also added to the problem by flooding the major rivers that drained onto the coast.

Global Weather Patterns

In discussing weather patterns along the coast, it is necessary to consider global wind systems. An idealized model of wind circulation on earth involves three large convection cells on each side of the equator. Although wind patterns at many places on the earth's surface differ significantly from the simplified pattern, it provides a good generalized picture of atmospheric wind flow.

An idealized cross-section through the atmosphere shows that pattern of the three global convection cells (Figure 2–23). Along the equator, the sun's rays pass straight through the atmosphere, heating the surface of the continents and oceans. Hot, moist air converges at the equator and rises vertically, moving north and south toward poles in the upper atmosphere. The hot oppressive air along the equator is called the *doldrums* from the early days of sailing ships.

FIGURE 2–23. Idealized global circulation proposed for a three-cell circulation model.

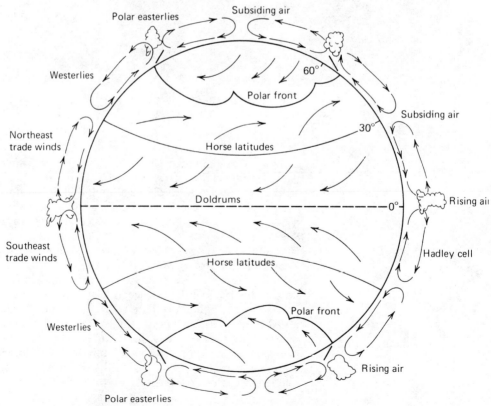

As the air moving poleward from the equator is cooled, it descends in the horse latitudes, about 30 degrees north and south of the equator (Figure 2–23). The descending air is more dense and forms a subtropical high-pressure area of low humidity and clear skies. The horse latitudes were named for horses floating in the ocean when they were thrown overboard by becalmed sailing ships.

The easterly trade winds blow from the northeast and southeast in belts between the doldrums and the horse latitudes (Figure 2–23). The trade winds are steady dry winds, with an average speed of about 15 knots. The trade winds north of the equator are strongest during the summer when the subtropical high is strongest.

To the north of the horse latitudes, a second convection cell is formed with the prevailing westerlies blowing toward the poles. The winds are deflected toward the right, producing southwest winds in the Northern Hemisphere (Figure 2–23). Where the westerlies blow over water, they pick up moisture, providing moist, warm air typical of a humid summer day. The westerlies blow almost entirely over oceans in the Southern Hemisphere. They have become known as the roaring forties, furious fifties, and screaming sixties for the latitudes at which these strong winds occur. In the Northern Hemisphere, continental masses are more extensive and the prevailing westerlies interact with the polar easterlies to produce low-pressure systems and their associated fronts.

Monsoon winds reverse direction with the season. The best known monsoons occur in India, where winter winds blow from the northeast. These cool, dry winds originate in Asia. In summer, the trade winds and westerlies combine to form the southwest Himalayan monsoons. The hot humid air from the Indian Ocean rises when it hits the mountains, unleashing the rainy season of the Indian summer.

The polar front forms the northern boundary of the westerlies in the Northern Hemisphere (Figure 2–23). The polar front is a belt of contact between warm, moist air brought up from the south by the westerlies and cold, dry air from the arctic. Frequent storms are generated along the polar front where warm, moist, light air is forced over cold, dry, dense air.

Coastal Climates

Weather includes the daily, weekly, and seasonal changes in air temperature, humidity, precipitation, wind speed and direction, and cloud cover, which affect the coast. Climate is considered to be the average weather for an area over an interval of at least 30 years. Although weather in an area will vary considerably over a year, the climate remains fairly constant for the short term.

Five major climatic types cover most of the world's coasts. First, the temperate maritime climate affects the west coast of

Europe and North America. Second, the climate of continental interiors influences the east coasts of continents including North America and eastern Asia. Third, the monsoon climate of southern Asia has a seasonal reversal in wind direction and rainfall. Fourth, the subtropical rainy and dry zones are under the influence of the trade winds, and fifth, the equatorial rainy zone occurs along the equator.

Temperate maritime climate along western edges of continents is particularly strong in western Europe, western North America, the southern parts of western South America, New Zealand, and Australia. In North and South America, long mountain ranges parallel to the edge of the continent confine the maritime climate to a narrow belt along the coast. However, the broad lowlands of western Europe are under the influence of the maritime climate.

The maritime climates are characterized by mild winters and moderately warm summers. The rainfall is concentrated in the fall and winter months, with fog common along the coast during the summer. In Washington and Oregon, the rainfall of 100 to 200 centimeters (40 to 80 inches) per year is concentrated in the four months from November through February. During summer, wind circulating in a clockwise motion around the North Pacific high produces upwelling and fog along the Pacific coast. In San Francisco, the fog bank creeps in beneath the Golden Gate Bridge. In Los Angeles, the coastal fog is often trapped by a temperature inversion and mixed with industrial and automobile emissions to produce smog. The smog is trapped in the Los Angeles Basin by the surrounding mountains.

The east coasts of North and South America are influenced by continental air masses. Along the polar front, continental polar and arctic air masses interact with maritime tropical air masses to produce extremes in temperature. Tattosh Island off the west coast of Washington has a mean temperature range of 5 degrees Centigrade for the coldest and warmest months. At the same latitude on the east coast of North America, Saint Johns, Newfoundland, has a range of 21 degrees Centigrade. In Europe and Asia, the contrast is just as striking. Brest, France, on the west coast of Europe has a range of 9.9 degrees Centigrade between the warmest and coldest months, while Otiai, Sakhlin, at the same latitude on the east coast of Asia has a range of 32.5 degrees Centigrade.

The eastern coasts of North America and Asia are buffeted by thunderstorms in the summer, blizzards in the winter, and hurricanes in the late summer and fall. Warm, moist air from the Gulf of Mexico supplies the moisture that provides a steady parade of storms across the northeastern tier of states.

The monsoon climate of southern Asia has a reversing dominant wind direction in winter and summer. On the coast of India, the northeast monsoon lasts from January to May, and the southwest monsoon extends from June through December.

January is the most pleasant month in India, with northwesterly winds descending from the Himalaya Mountains with little or no rain. In June, the southwest monsoon arrives with a warm moist air mass from the Indian Ocean. June through September is the season of coastal rains, although there are often breaks with clear skies for several days or even weeks. In October and November, the southwest monsoons slowly retreat as the mountain air makes its way south.

The trade wind climates in the subtropics are generally dry with steady winds on the eastern side of continents, and seasonal winds on the western margins. The subtropical winter rainy zone occurs somewhat farther from the equator (30–40° north and 30–35° south). In summer, the area is under the influence of the dry trade winds, producing fine dry weather. However, in winter, the prevailing westerlies take over, and cloud cover and precipitation affect the area. The subtropical zone is also the spawning ground for the tropical cyclones that interrupt the peaceful trade winds with powerful hurricanes.

The equatorial rainy zone is the area of the doldrums. Hot humid days and nights are punctuated by afternoon thundershowers that don't relieve the heat, but just add more moisture. Commodore Arthur Sinclair crossed the doldrums on the frigate *Congress* in 1819, and described it as follows:

> No person who has not crossed this region can form an adequate idea of its unpleasant effects. You feel a degree of lassitude unconquerable, which not even sea bathing, which everywhere else proves so salutary and renovating can dispel. Except when in actual danger of shipwrecks, I never spent twelve more disagreeable days in the professional part of my life than in these calm latitudes.

Summary

Coastal weather and climate are influenced by ocean waves and currents. The ocean heats up and cools off more slowly than the land, resulting in cooler summers and warmer winters along the coast. Clouds, fog, and rain are produced when the warm, moist maritime air moves over land. Clouds and changes in wind direction are useful for short-term weather predictions along a coast.

Winds on the earth's surface result from differences in barometric pressure. Air moving from an area of high pressure to one of low pressure is deflected to the right in the Northern Hemisphere by the Coriolis effect. A geostrophic wind flows in a counterclockwise direction around a low, clockwise around a high. Due to surface friction, wind is deflected into the center of a low and out from a high. Sea and land breezes are produced by differences in temperature over land and sea.

Coastal weather is influenced by air masses and related fronts. The cold, dry continental polar air mass interacts with the warm, moist maritime tropical air mass along the polar front. Most changes in weather are related to the passage of warm, cold, and occluded fronts.

Damage along the coast is due to wind, waves, and rain during coastal storms. High winds and rain occur during thunderstorms and squalls. Hurricanes originate in the tropics and move to the north or south, into the mid-latitudes, causing death and destruction.

Coastal climates are defined as the average weather over at least 30 years. The western edges of continents are influenced by the warm, moist maritime climate, while the eastern side is under the effect of cold, dry continental air with greater extremes in temperature.

Selected Readings

BARRY, R. G. AND R. J. CHORLEY. *Atmosphere, Weather and Climate* (3rd ed.). London: Methuen and Co., Ltd., 1976. An introduction to weather with a detailed section on climates.

BATTAN, L. J. *Fundamentals of Meteorology.* Englewood Cliffs, N.J.: Prentice-Hall, Inc., 1979. A well-illustrated introduction to meteorology for the general public.

COLE, F. W. *Introduction to Meteorology* (3rd ed.). New York: John Wiley and Sons, 1980. A college level text with good diagrams and photographs.

DONN, W. L. *Meteorology* (4th ed.). New York: McGraw-Hill Book Company, 1975. An introduction to meteorology with coastal applications, containing rules for weather forecasting based on clouds and winds.

EAGLEMAN, J. R. *Meteorology—the Atmosphere in Action.* New York: D. Van Nostrand and Company, 1980. A nonmathematical introduction to meteorology using satellite data and computer maps. An interesting section on hurricanes.

LUTGENS, F. K. AND E. J. TARBUCK. *The Atmosphere—An Introduction to Meteorology* (2nd ed.). Englewood Cliffs, N.J.: Prentice-Hall, Inc., 1982. An excellent general text on meteorology with abundant diagrams and illustrations.

NAVARRA, J. G. *Atmosphere, Weather and Climate—An Introduction to Meteorology.* Philadelphia: W. B. Saunders Company, 1979. A good explanation of fronts and weather maps.

WEISBURG, J. S. *Meteorology, the Earth and its Weather* (2nd ed.). Boston: Houghton-Mifflin Company, 1981. Good coverage of frontal systems and storms along the coast.

fishing off a pier   Cape May, New Jersey

THREE
Waves

Introduction To Waves

Most people have gained their experience and knowledge of waves by watching them break along the coast. On a quiet day, waves gently lap against the shore, slowly shifting the sand along the beach. During a storm, the waves break with explosive force along the coast, removing the sand from the beach and attacking the dunes or cliffs behind the beach. The material eroded from the cliffs is ground up into gravel or sand, which continually replenishes the sand lost from the beaches (Figure 3–1).

Most waves in the ocean are generated by the wind that blows across the surface of the water. When the wind starts to blow, small capillary waves are formed on the surface, which quickly die out due to surface tension. If the wind continues to blow, large waves develop. The size of the waves is limited by the speed of the wind, the duration or length of time for which the wind is blowing, and the fetch or distance of water over which the wind is blowing. Large storms at sea are responsible for most of the waves at the beach. When the waves move out of the storm area, they become swells, which have been known to travel half the distance around the world.

When waves leave deep water, and approach the coast, the size and shape of the waves change. Five basic processes, including (1) shoaling, (2) breaking, (3) reflection, (4) refraction, and (5) diffraction, are modifying the waves in shallow water. Wave shoaling encompasses the changes in the size, shape, and speed

FIGURE 3–1. A large swell breaking off the Oregon coast with spindrift blowing off the top of the wave. *Photo by W. T. Fox.*

of a wave as it moves from deep to shallow water. Wave breaking involves the dissipation of wave energy along the coast and the transfer of energy from the waves to the process of erosion. Breaking waves are also responsible for the generation of currents along the shore, which transport sand along the beach.

Wave reflection is the bouncing of wave energy off of a steep beach or seawall and the set up of standing waves parallel to the shore. Wave reflection is also influential in the formation of sand bars. Wave refraction involves the bending of the wave crest and focusing of wave energy as a wave approaches the shore or bends around a point of land.

Wave diffraction is the scattering of wave energy as it moves past an obstruction such as a point of land or the tip of a breakwater. Wave reflection and diffraction are critical factors in harbor design and engineering.

Description of Waves

Before discussing the generation of waves and the changes that occur in waves as they move from deep water to the shore, it is necessary to introduce a few definitions of wave properties (Figure 3–2).

Wave length *(L)* is the horizontal distance between a point on a wave and the corresponding point on the following wave. Wave length is usually measured from one crest to the next.

Wave height *(H)* is the vertical distance between the top of the crest and the bottom of the trough in a wave.

The amplitude *(A)* is one-half the wave height which is also the vertical distance that the wave moves above or below the still water level.

The period *(T)* of a wave is the time it takes for a complete wave to pass a point. The period is usually measured in seconds.

Wave frequency *(F)* is the number of waves, or fraction of a wave, per second and is given in units of hertz. For example, a wave with a period of 5 seconds has a frequency of ⅕ or .2 hertz.

The speed at which the wave crest moves forward through the water is known as the phase velocity *(C)*. The wave energy

FIGURE 3–2. Symmetrical profile of a wave in deep water.
Drawing by C. W. L.

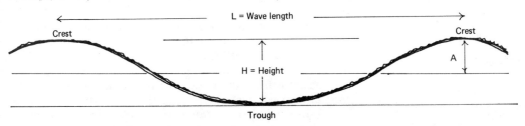

moves forward at the group velocity *(Cgp)*. Water particles within the wave are moving in circular or elliptical orbits at the particle velocity *(V)*. Phase velocity is important in the analysis of wave refraction, and group velocity is used in determining wave height as waves enter shallow water.

The shape of a wave in profile and along the crest varies with the water depth. In deep water, a wave has a symmetrical profile and the cross-section of a wave resembles a sine curve with a smooth crest and trough (Figure 3–2). In shallow water, a wave has an asymmetrical profile with a peaked crest and a broad trough. In deep water, the length along the crest is short relative to the wave length, and in shallow water, the crest length is long relative to the wave length.

Wave length   The length of waves varies considerably and is usually quite difficult to measure. When the ocean waves are approaching directly onshore, it is possible to use a fishing pier for measuring wave length. Wave length can also be measured from aerial photos or satellite images of the sea surface. However, in most coastal studies, no attempt is made to measure wave length directly. Instead it is usually derived from the wave period.

Waves from storms on the North Atlantic usually have an average wave length of about 100 to 150 meters (325 to 500 feet). In the Pacific, the average lengths of storm waves vary from about 180 to 200 meters (600 to 650 feet). In the Antarctic Ocean, heavy gales in the storm belt may persist for several days and produce wave lengths of 200 to 250 meters (650 to 800 feet).

When waves move out of the storm area where they are generated, longer waves move ahead of the storm and form swells. In swells, smooth profile lengths of more than 300 meters (1000 feet) are quite common on the open ocean. On the Atlantic, the maximum recorded swell length is about 760 meters in the north and about 820 meters in the equatorial Atlantic. Swells in the Pacific are often more than 1000 meters long.

Wave height   The height of a wave is much easier to measure than the length; however, it is also common to overestimate the height of a wave. When you hear reports from untrained observers of giant wave heights, take the reports with a grain of salt. Wave estimates, like crowd estimates at a parade, tend to be grossly exaggerated.

When the waves are small, it is possible to measure wave heights in the surf zone with a calibrated staff. Markings are painted along the staff at 2-centimeter or one-tenth-of-a-foot intervals. Wave heights are measured directly by reading the vertical distance from the bottom of the trough to the top of the crest of the next wave.

When the wave height is greater than about 1 meter, it is difficult to measure wave height directly in the surf zone. For larger waves a fishing pier with open pilings is useful for measuring wave height. In this case, a plank of wood painted white with clear markings at one-tenth-of-a-meter intervals is bolted to one of the pilings of the pier. The plank should be placed near the deep end of the pier where it is easy to read from the walkway.

Electronic devices have been installed on fishing piers so that wave height can be measured on the pier and recorded on the shore. The step-resistance wave staff has a series of brass screws that project from the wave staff at 3-centimeter intervals. A weak electrical current is passed through the staff from the shore. As the screws are submerged by a wave, the resistance changes and the water level in the wave is recorded on shore. By taking a series of readings, a fraction of a second apart, it is possible to record the size and shape of a wave. At several coastal locations wave recorders are turned on for 20 minutes, four times each day, to record the average wave height and period.

It is also possible to record wave height and period using a pressure sensor, which is placed on the bottom of the ocean. The sensor measures the water pressure over the sensor, which can be converted to wave height. Since the pressure is measured in a cone over the sensor, it is best not to place the sensor too deep or it will superimpose the pressure changes from more than one wave. Usually, a pressure sensor is placed at a depth of 10 to 15 meters, where it will measure individual waves but will not be buried by shifting sand. A breaking wave 10 meters high was recorded with a pressure sensor at Monterey, California (Figure 3–3). A float, one-half meter in diameter, which was tethered to the meter, can be seen in the face of the breaking wave.

FIGURE 3–3. A 10-meter high breaking wave at Monterey, California. The vertical bar is the same height as a person standing in the wave.
Photo by Peter Howd, Oregon State University.

A microseismograph has been successfully used at the Oregon State University Marine Science Center for measuring wave height and period. A seismograph is usually used for recording earthquake waves that pass through the earth. A microseismograph is an extremely sensitive instrument that records small vibrations in the surface layers of rock. As large waves move across the surf zone, vibrations are transmitted to the seismograph on shore. During severe winter storms, waves up to 15 meters high have been recorded by the seismograph, waves that would have buried or destroyed almost any other type of wave recorder.

Although occasional waves of 10 to 15 meters are reported, they are not very common, and even in the worst storms, most of the waves are considerably smaller. Based on more than 40,000 extracts taken from ship logs, it has been estimated that 40 percent of the ocean waves are less than 1 meter high, 80 percent are less than 4 meters, and only 10 percent are greater than 6 meters.

Keeping in mind that exceptionally large waves are rare events, it is interesting to look at some of the reports of high waves. The largest waves reliably reported at sea were encountered by a tanker, the U.S.S. *Ramapo*. The *Ramapo* was proceeding from Manila to San Diego on February 7, 1933, when exceptionally large waves were encountered. Lieutenant Marggraff was standing watch on the bridge when he saw seas astern at a level above a boom against the crow's nest or lookout platform. At that moment of observation, the horizon was hidden from view by the waves that were approaching from the stern (Figure 3–4). By knowing the length of the ship and the height of the crow's nest, he estimated that the height of the wave was at least 34.1 meters (112 feet).

During severe storms, rocks have been thrown through the glass of the Tillamook Rock Lighthouse, 40 meters (133 feet) above the ocean near the Oregon coast. The lighthouse is located on a rock about 2 kilometers offshore. The water surrounding the rock is between 20 and 40 meters deep. The heights of the waves that damaged the light were increased by the relatively

FIGURE 3–4. The U.S.S. *Ramapo* encountered a 34.1 meter (112-foot) wave on February 7, 1933, in the South Pacific. The height was estimated by sighting from the bridge over the crow's nest to the wave crest.
Drawing by C. W. L.

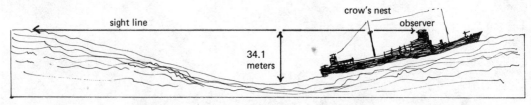

sight line

crow's nest

observer

34.1 meters

shallow depth. Therefore, the 40-meter wave at Tillamook Light was a shallow water wave and does not represent a maximum height for deep water waves, a record still held by the U.S.S. *Ramapo*.

In order to have some idea about the heights of 10-, 20-, or 40-meter waves, it is a good idea to compare the heights with some familiar objects. The average height of a single story in a building is about 3.5 meters, so a 10.5-meter wave would be equivalent to a 3-story building and a 35-meter wave would be as high as a 10-story building. The height of a large mature tree is about 20 meters, and a church steeple may reach 30 meters. This may give you some respect for Lieutenant Marggraff, who had the courage to record his sightings while 30-meter waves were bearing down on his ship.

Wave period and frequency

Of all the properties of waves, the period is the easiest to measure. The period can be measured from the beach by timing with a stopwatch the arrival of ten successive waves. In deep water, the period can be determined by timing the rise and fall of a small floating object such as a piece of seaweed.

Since period is the time interval between successive crests, one usually counts eleven wave crests to get ten wave intervals. To get the average wave period, divide the time interval for ten waves by ten. For example, if the time interval for ten waves is 67 seconds, the average wave period would be 6.7 seconds. Many coastal engineers prefer wave frequency, which is one over the wave period $(F = 1/T)$.

Periods for waves in the ocean vary from less than one-tenth of a second to more than 24 hours. Waves with periods of less than one-tenth of a second are considered capillary waves. Capillary waves respond to the surface tension as the main restoring force. The ionic bonds that hold molecules together within the water create a thin skin on the surface, which produces surface tension. It is possible to see surface tension when you float a needle on a glass of water. The water bends up forming a meniscus around the edge where it comes in contact with the glass, and bends down along the needle due to the weight of the needle. If you disturb water in a glass, it will quickly become smooth again as surface tension restores the flat surface. Capillary waves generated by the wind also disappear quickly when the wind stops.

Waves with periods greater than one-tenth of a second and less than 5 minutes are considered gravity waves. Gravity waves include the familiar wind-driven waves that break against the coast. The majority of gravity waves have periods between 5 and 20 seconds. The main restoring force for gravity waves is the mass of the water in the wave responding to the force of gravity.

The waves with periods between 30 seconds and 5 minutes are generally stationary or standing waves, which are produced by the reflection of gravity waves off the coast. Standing waves include edge waves, which travel parallel to the shore, and surf beat, which moves on and off the shore.

Long period waves have periods between 5 minutes and 12 hours. The long period waves are often the result of storms or rapid changes in barometric pressure, which cause a rise or fall in sea level. A seiche is a wave set up by a sudden change in barometric pressure on a lake or large bay. During a seiche, water sloshes back and forth across the body of water forming a standing wave. The period of the seiche is determined by the size and shape of the body of water.

A submarine landslide or an earthquake can form a long period wave, or tsunami, which can travel thousands of kilometers. The tsunami usually has a period of about 15 minutes and a length of several hundred kilometers.

The tides with periods between 12 and 24 hours are standing waves with a rotational motion due to the Coriolis effect. Tide waves are set up in the oceans by the pull of gravity exerted by the sun and the moon. The forces and behaviors of tides are explained in Chapter 4.

Waves with periods greater than 24 hours are due to changes in the position of the moon in relation to the earth during the lunar month and during the year. Other long-period waves are caused by seasonal differences in barometric pressure over different parts of the ocean.

Wave shape

Wave steepness, the ratio of wave height to wave length (H/L), is an important aspect of waves, for both boat handling and for beach erosion. Storm waves with high steepness are relatively short, high, and choppy, while swells with low steepness are long, low, and rounded (Figure 3–5). Small waves that are relatively steep can make life very uncomfortable in a small boat. The bow will plunge into the trough of a wave while the stern is lifted up by the crest. If a boat is several times longer than the wave length, the boat will be much less affected by short, steep waves. However, even in a large ship, long low swells moving across the ship can cause a sideways roll that will bring on seasickness.

Steep waves that accompany storms cause erosion of sea cliffs and sand dunes along the coast. Swells with low steepness values generally result in the shoreward transport of sediment. Therefore, in the quiet interval between storms, sediment that was carried offshore by the storm waves is returned to the beach.

Wave motion

It is interesting to study the motion of water particles in a wave, and to see how this motion changes as a wave moves from deep water toward the shore. It is impossible to follow an individual particle of water in a wave, but it is possible to follow the water motion by throwing a small floating object such as a cork into the wave (Figure 3–6). As the crest approaches, the cork rises

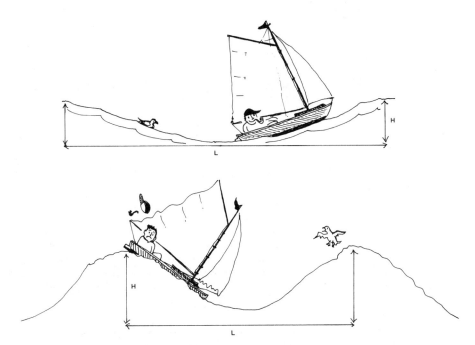

FIGURE 3–5. A small boat and its skipper react differently to waves of low steepness (top) and waves of high steepness (bottom).
Drawing by C. W. L.

FIGURE 3–6. In deep water, the water particles in a wave move in circular orbits as shown by the cork. The diameter of the orbit decreases with depth. The wave is moving to the right.
Drawing by C. W. L.

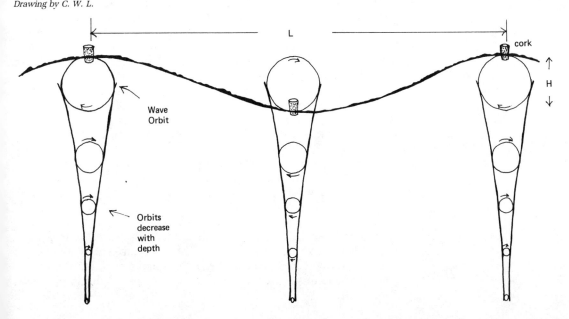

and moves forward with the crest. After the crest has passed, the cork drops into the trough and moves backward in the trough. Therefore, as the wave passes, the cork follows a circular path that is known as the wave orbit. The diameter of the orbit is equal to the height of the wave. The orbits are not perfect circles and the shoreward component of the water motion is called mass transport.

Next, consider how orbits change with depth. At the surface, the diameter of the orbit is equal to the height of the wave (Figure 3–6). As a rule of thumb, the diameter of the orbit is cut in half for each one-ninth of the wave length's increase in depth. Therefore, if a wave is 90 meters long and four meters high in deep water, the diameter of the orbit will be 2 meters at a depth of 10 meters, one meter at a depth of 20 meters, one-half meter at 30 meters, and 25 centimeters at 40 meters. In going to the depth equal to one-half of the wave length, the diameter of the orbit decreases to about .05 of the wave height. For this reason, submarines can travel 50 meters beneath the surface during a raging storm and not be affected by the waves.

Deep-Water Waves

Most waves that strike the coast were formed by storms in the open ocean. In deep water, where the depth of the body of water is greater than half the length of the wave, the wave is unaffected by the bottom. As the wave approaches the coast, it enters shallow water, where the depth is less than half the wave length. In shallow water, the bottom interferes with the orbital motion of the water particles and distorts the shape of the wave.

Wave length is the critical factor in deciding whether a wave is in deep or shallow water. The lengths of waves vary from a few centimeters for capillary waves, to hundreds of meters for storm waves and swells, and several kilometers for tsunamis that were generated by earthquakes. A storm wave with a length of 100 meters would be a deep-water wave for any depth over 50 meters. On the other hand, a tsunami with a length of 20 kilometers would be a shallow-water wave over the deepest parts of the ocean.

Most storm-generated waves travel great distances over deep water as swells before they arrive at the coast. We will consider the generation of storm waves in deep water, and their transformation into swells as they move out of the storm area.

Wave generation

If the surface of a pond is glassy smooth, a puff of wind will form a cat's paw. The cat's paw is a roughly oval area covered by a diamond-shaped pattern of capillary waves. Capillary waves are small waves that form in the thin surface layer of the water. Surface tension forms a thin skin across the water. When the wind blows, the surface of the water vibrates like the head of a drum. Capillary waves grow to a maximum length of about 1.73 centimeters. If the breeze stops, capillary waves die out quickly due to surface tension and the cat's paw will disappear.

If the wind continues to blow, capillary waves grow and are transformed into gravity waves. One of the important differences between capillary and gravity waves is the durability of the gravity waves. Capillary waves rapidly disappear, but gravity waves continue to expand until they eventually die out or break on shore.

Storm waves   When a large storm blows across part of the ocean, waves of many different sizes and shapes are generated beneath the storm. The storm waves cover a wide spectrum of wave lengths, heights, and periods (Figure 3–7). From a statistical analysis of the wave spectrum, it is possible to compute most frequent wave height and period, average wave height and period, and significant wave height and period. The wave spectrum generally includes a large number of small waves and a small number of very large waves. Therefore, the most frequent wave height and period are less than the average wave height and period.

The significant wave height is the average of largest one-third of the waves and significant period is the period that accompanies significant wave height. The significant wave height

FIGURE 3–7. Averages of selected wave spectra for winds from 37 to 74 km/hour (20 to 40 knots).

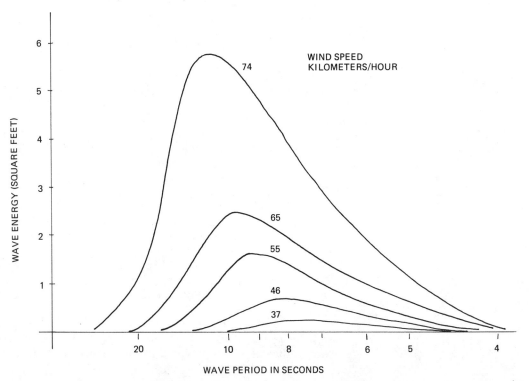

is used by coastal engineers and designers for buildings and structures that must survive in the surf zone.

Wave spectrum. The wave spectrum is determined by wind speed, duration, and fetch. The wind speed, which is measured in miles per hour or knots, is directly related to the Beaufort Force (Table 2–2). The relationship between Beaufort Force and sea state is vividly given in Table 2–4. As wind speed increases from a breeze through a gale to a hurricane, waves increase from small ripples through large waves covered with spray and foam. Wind duration is the length of time that wind of a particular speed is blowing over the ocean. The fetch is the distance over which the wind is blowing to generate storm waves.

The most frequent, average, and significant wave heights can be estimated from the wind speed and duration. For example, if the wind blew over the open ocean at 25 knots for 4.5 hours, the most frequent wave height would be .85 meters (2.8 feet), the average wave height would be 1.1 meters (3.6 feet) and the significant wave height would be 1.7 meters (5.6 feet). The significant wave period under these conditions would be about 6 seconds.

If the wind speed is increased or if the wind blows for a longer duration, the sizes of waves in the spectrum are increased. For example, if the wind blew at 45 knots for 5 hours, the most frequent wave height would be 2.5 meters (8.2 feet), the average wave height would be 3.2 meters (10.5 feet), and the significant wave height would be 6.4 meters (21 feet). The significant wave period would be 10 seconds.

Fully developed sea. If the wind continues to blow for an extended period of time, a fully developed sea state is formed (Figure 3–8). The surface of the sea reaches a theoretical equilibrium and the waves do not continue to grow. It usually takes 3 to 5 days to reach the maximum sea state. However, 90 percent of the maximum sea state, which is considered a fully developed sea, can be reached in 18 to 25 hours. For example, if the wind blows continuously for 20 hours at 30 knots, the average wave height would be 5 meters (16 feet) and the significant wave height would be 6.1 meters (20 feet) for a fully developed sea. However, if the wind blows at 60 knots, a fully developed sea is reached at 25 hours when the average wave height is about 20 meters (66 feet) and the significant wave height is about 27 meters (88 feet). The record waves encountered by the U.S.S. *Ramapo* (34.1 meters or 112 feet) were generated by sustained high winds, which blew for several days in the South Pacific.

Wave fetch. Wave fetch is the horizontal distance on the surface of the ocean over which the wind blows. Fetch distance

FIGURE 3–8. Storm waves eroding the beach and bluff at
Stevensville, Michigan.
Photo by W. T. Fox.

is determined by both the size of the storm and the size of the
body of water. For a large body of water such as the ocean,
fetch is controlled by the diameter of the storm. However, for a
small body of water such as a lake or a bay, fetch is limited by
the minimum distance in the path of the storm.

A certain minimum fetch is required for waves to reach a
fully developed sea state. For a wind of 20 knots, fetch must be
at least 333 kilometers (200 miles) to reach a fully developed
sea. When wind speed is increased to 40 knots, a fetch of at least
830 kilometers (500 miles) is required to reach a fully developed
sea. At wind speeds of 60 knots, fetch must be at least 1280 kilometers (800 miles) for a fully developed sea.

Swells When waves leave the storm area where they were generated,
they become swells, which are sorted into different lengths,
heights, and periods. In the storm area, the wave spectrum includes a wide range of wave lengths, heights, and periods. In
deep water, the speed of the wave or phase velocity (C_o) can be
computed directly from the wave period (T).

Phase velocity $C_o = 5.12 \ T$ (feet/second)
$C_o = 1.56 \ T$ (meters/second)

The waves with longer periods travel at a faster speed. For example, if a wave has a period of 5 seconds, the phase velocity

would be 7.8 meters per second (25.6 feet per second). A wave with a period of 10 seconds would have a phase velocity of 15.6 meters per second (51.2 feet per second). Therefore, the wave with a period of 10 seconds would be traveling twice as fast and would move out ahead of the 5-second wave. As the waves spread out from the storm area, they are sorted according to wave period, with the longer-period waves out front and the shorter-period waves trailing behind.

Wave length in deep water (L_o) is directly proportional to the square of the period.

$$\text{Wave Length} \qquad L_o = 5.12 \ T^2 \ \text{(feet)}$$
$$L_o = 1.56 \ T^2 \ \text{(meters)}$$

For example, a wave with a period of 5 seconds would have a length of 39 meters (128 feet) and a wave with a period of 10 seconds would have a length of 156 meters (512 feet) in deep water.

The chaotic sea state that exists in the storm area is sorted out as the swells move across the ocean toward a distant shore. The first waves to arrive at the shore are low-long-period swells. These are followed by higher swells with shorter periods. As the swells extend out from the storm, they form ever-expanding rings of waves similar to those formed when you drop a pebble in a pool of water. As the rings enlarge, the wave is spread out along a greater length, the expanding circumference of the circle. Therefore, the wave height gradually decreases as the swells move away from the storm area. When the swells arrive at the coast, they form a uniform succession of waves with about the same period and height. Of course, the period and height changes with time as the slower swells arrive at the coast.

Shallow-Water Waves

When waves move from deep to shallow water, several processes that change the size and shape of the waves are operating. As long as the water depth is greater than one-half the wave length, the waves are considered deep-water waves. At depths of less than one-half the wave length, wave height, length, and phase velocity are changed by shoaling or interaction between the wave motion and the bottom. When waves approach the shore, wave shape is distorted and waves break along the beach. If the shore is relatively steep, waves are reflected back out to sea. When waves approach the shore at an angle to the beach, wave crests are bent by refraction. Finally, when waves pass the end of a point of land or the tip of a breakwater, a circular wave pattern is generated behind the breakwater by diffraction. The diffracted wave pattern intersects the incoming waves, resulting in increased wave heights at the entrance to a harbor.

The shallow-water wave processes, including shoaling, breaking, reflection, refraction, and diffraction, influence erosion patterns along the coast. The beaches and barrier islands that will be considered in the following chapters are largely controlled by shallow-water wave processes.

Shoaling waves

When shoaling waves approach a beach, several characteristics of the waves change. The waves slow down as they start to "feel bottom," resulting in a decrease in speed or phase velocity, C. When waves slow down, the distance between waves or wave length, L, becomes less. Although the speed and length of the waves decrease, wave period, T, or time interval between waves, does not change. If the period changed, it would mean that more waves were formed between the earlier waves. On the other hand, wave height acts in a rather peculiar manner. As the wave first enters shallow water, wave height starts to decrease. However, as the wave moves through the surf zone and approaches the beach, wave height increases until the wave becomes unstable and breaks (Figure 3–9).

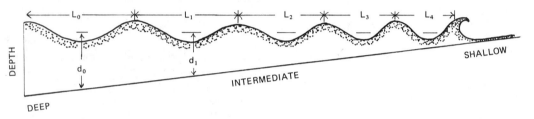

FIGURE 3–9. When a shoaling wave moves from deep water (d_0) into shallow water (d_1). The wave length (L_0) decreases (L_2 to L_4) and the wave height first decreases, then increases as the wave breaks.
Drawing by C. W. L.

The changes that take place in length, height, and phase velocity of shoaling waves are explained by small amplitude wave theory. The theory assumes that waves have very small amplitude or height relative to their length. Wave equations based on the law of continuity and Newton's second law have been developed using sophisticated mathematical arguments to explain water motion in small amplitude waves. It is beyond the scope of this book to go into the mathematics involved in wave theory. For those who are interested in the physics of waves and the development of the mathematical theory, I would strongly recommend the book *Wind Waves* by Blair Kinsman, which is written by an oceanographer and gives a very clear and readable explanation of wave theory.

The ratio between depth of water, d, and wave length in deep water, L_0, is referred to as relative depth (d/L_0). When relative depth is greater than ½ $(d/L_0 > ½)$, it is considered deep water. When relative depth is between ½ and ¹⁄₂₅, it is considered intermediate, $(½ > d/L_0 > ¹⁄₂₅)$. When relative depth is less than ¹⁄₂₅ $(d/L_0 < ¹⁄₂₅)$, it is known as shallow water. The expressions for calculating phase velocity, C, wave length, L, and wave height, H, based on depth, d, and wave period, T, are given in Table 3–1.

Table 3–1. CHANGES IN PHASE VELOCITY, WAVE LENGTH, AND WAVE HEIGHT IN SHOALING WAVES.

WATER DEPTH	RELATIVE DEPTH	PHASE VELOCITY	WAVE LENGTH	WAVE HEIGHT
Deep	$d/L_o > 1/2$	$C_o = 1.56T$	$L_o = 1.56T^2$	H_o
Intermediate	$1/2 > d/L_o > 1/25$	Decreasing	Decreasing	Decrease to $d/L_o = 1/6$, then increases
Shallow	$1/25 > L/d_o$	$C = \sqrt{gd}$	$C = \sqrt{gd} \cdot T$	increase $> H_o$

The concept of relative depth is important in the study of waves. Small waves change size and shape in the same way as large waves when they move from deep to intermediate and shallow water. Therefore, it is possible to study small waves in a scale-model wave tank and apply the results to large waves in the ocean. For example, a small wave with a length of 10 meters would be considered a deep-water wave as long as the depth is greater than 5 meters $(d/L_o > 1/2)$. The 10-meter wave would become a shallow-water wave when the depth is less than 0.4 meters (40 centimeters or 17 inches). On the other hand, a large ocean swell with a length of 1000 meters would be a deep-water wave for all depths greater than 500 meters. The 1000-meter wave would become a shallow-water wave when the depth becomes less than 40 meters (131 feet).

As waves move into shallow water, the shapes of the orbits change from circular to elliptical or oval (Figure 3–10). In shallow water, the orbits look as if they have been squeezed from above. Near the bottom, the orbits become flattened until they are almost horizontal.

FIGURE 3–10. In shallow water waves, orbits are elliptical near the surface and become horizontal near the bottom. Drawing by C. W. L.

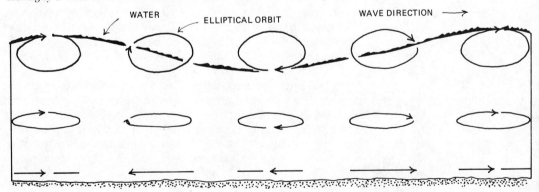

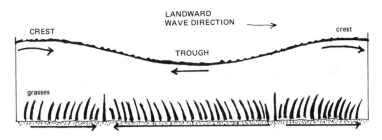

FIGURE 3–11.
Grass growing on the bottom in shallow water bends with the direction of water motion in the wave, landward under the crest and seaward under the trough.
Drawing by C. W. L.

It is possible to observe the motion of shallow-water waves with a face mask near a beach. It is best to observe waves when the water is clear, the waves are about one-half meter high, and water is about a meter deep. As a wave crest sweeps by, sand is carried seaward. Grass or seaweed growing along the bottom will bend forward under the crest of the wave, and bend out to sea under the trough (Figure 3–11). When you stand in the surf zone, you will feel yourself being pulled toward the breaking wave as the wave approaches. As the wave breaks, you are carried toward the shore along the crest of the wave.

Breaking waves

When a wave reaches the coast, it either breaks along the shore or is reflected back out to sea. The behavior of the wave at the shore is largely determined by wave steepness and slope of the bottom near the beach.

Storm waves with a high steepness ratio are characterized by spilling breakers, which break over a wide section of the surf zone. (Figure 3–12). On steep beaches, storm waves may form plunging breakers against the beach.

Swells that were generated by distant storms will form spilling, plunging, collapsing, or surging breakers, depending on the bottom slope. If the slope is relatively flat, the wave will break, forming a spilling breaker (Figure 3–12). A spilling breaker is an over-steepened wave that starts to break at the

FIGURE 3–12.
Plunging, spilling, and surging breakers develop along the shore. Breaker type depends on bottom slope and wave steepness.
Drawing by C. W. L.

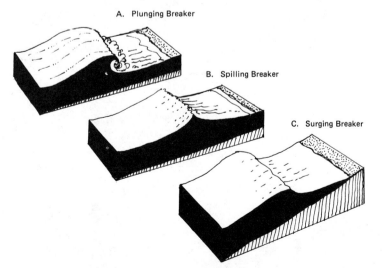

81

FIGURE 3–13. Spilling breakers form on the low slope off
Padre Island, Texas.
Photo by W. T. Fox.

crest and continues to break as the wave travels toward the
beach (Figure 3–13). A spilling breaker forms when the slope of
the bottom is less than 3 degrees. When the wave starts to spill,
the ratio of water depth to wave height is about 1.2 to 1. Spilling
breakers are the most common type and provide good waves for
surfing. The wave energy in a spilling breaker is dissipated over
a broad area of the surf zone.

A plunging breaker forms when the bottom slope is between 3 and 11 degrees. In a plunging breaker, the crest curls
over a pocket of air, forming a tube (Figures 3–12 and 3–14). As
the wave breaks, the tube moves toward the bottom, beneath
the wave, and stirs up bottom sediment. If the bottom slope is
between 3 and 7 degrees, the wave will break once offshore,
then reform and break again at the beach. If the slope is between 7 and 11 degrees, the wave will plunge once at the shoreline and wash up on the beach. Plunging breakers are the most
dramatic breakers, and do the most damage, because wave energy is concentrated where the wave breaks. Often a fountain is
seen behind a plunging breaker where the tube breaks through
the water surface behind the crest. The ratio of water depth to
wave height in a plunging breaker is 0.9 to 1.

In a collapsing breaker, the breaking is confined to the
lower half of the wave. A small air pocket forms as the wave

FIGURE 3–14. A plunging breaker on the steep foreshore at
Plum Island, Massachusetts.
Photo by W. T. Fox.

moves toward the beach, but most of the wave is reflected off
the beach. For a collapsing breaker, bottom slope is relatively
steep, between 11 and 15 degrees. The ratio of water depth to
wave height is about 0.8 to 1.

A surging wave develops on a steep bottom where the
slope is greater than 15 degrees. The wave does not break, but
surges up the beach face (Figure 3–12). A surging wave is reflected off the beach, generating standing waves near the shore.

Wave reflection   When waves approach a steep shoreline or a vertical sea wall, a
large portion of the wave energy is reflected. A reflected wave is
formed that has the same period as the incoming wave, but is
traveling away from the shore. When the reflected wave meets
the next incoming wave, a standing wave pattern is established
along the shore. Standing waves are important in the formation
of offshore bars, beach cusps, and rip currents, so it is worth a
little time studying the formation of standing waves.

If an incoming wave moves to the right and is reflected off
a vertical wall, a reflected wave moves from the wall to the left
(Figure 3–15). When the crest of the incoming wave is at the
wall, the incoming and reflected waves are in phase. At that
time, crests and troughs are superimposed and the resultant
wave height is double the incoming wave height (Figure 3–15A).

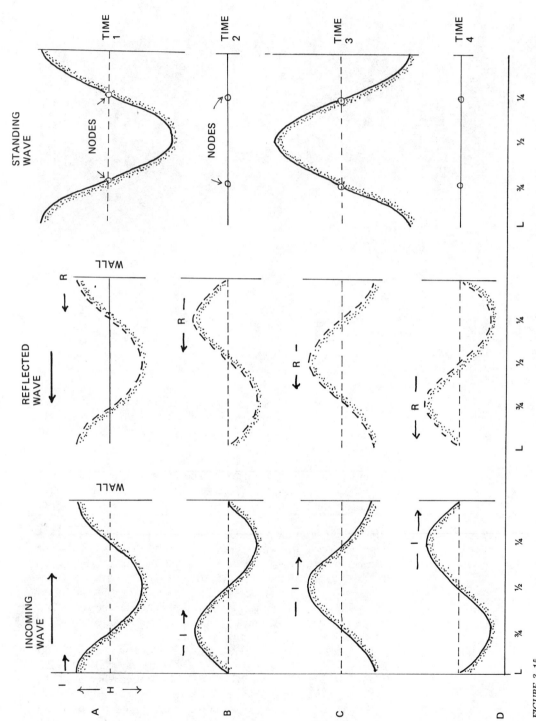

FIGURE 3-15.

A standing wave develops when an incoming wave (*I*) is reflected off a wall, forming a reflected wave (*R*), which moves to the left. The standing wave is formed by adding together the heights of the incoming and reflected waves.

Drawing by C. W. L.

In Figure 3–15B, the incoming wave has moved toward the wall and the reflected wave has moved away from the wall, one-forth of a wave length. At that time, the crest of the incoming wave coincides with the trough of the reflected wave and the waves are out of phase. The crest of the incoming wave is canceled out by the trough of the reflected wave and the water surface is flat.

In Figure 3–15C, the incoming wave has moved one-half wave length to the right and the reflected wave has moved one-half wave length to the left. Once again, crests and troughs of the incoming and reflected waves are superimposed and the combined wave height is double the incoming wave height. In Figure 3–15D, the incoming wave has moved three-fourths of the way to the right, and the crest of the incoming wave overlies the trough of the reflected wave. The waves cancel each other out, producing a flat water surface.

If the resultant waves from Figures 3–15A through D are superimposed, a standing wave pattern emerges (Figure 3–16). Nodal points are located at one-fourth and three-fourths wave lengths from the wall, where the water surface remains level. Antinodes are located along the wall, at one-half and 1 wave length from the wall, where the height of the standing wave is double the incoming wave height.

Next, consider the motion of the water in a standing wave. In a progressive wave, the water follows circular orbits in deep water and elliptical paths in shallow water. In a standing wave, on the other hand, the water has horizontal motion beneath the nodes and vertical motion under the antinodes (Figure 3–17). Under nodal points, water moves back and forth from a crest

FIGURE 3–16. A standing wave has nodal points with no vertical motion at distances of ¼ and ¾ wave lengths from the wall. Antinodes with double the incoming wave heights are located at the wall, and ½ and 1 wave length from the wall. *Drawing by C. W. L.*

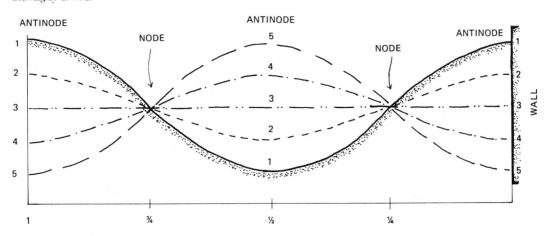

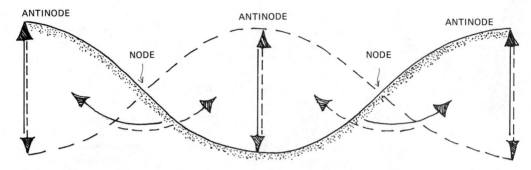

ANTINODE   ANTINODE   ANTINODE

NODE   NODE

FIGURE 3–17. The water motion is horizontal beneath the nodes and vertical beneath the antinodes of a standing wave. *Drawing by C. W. L.*

on one side of the node to the crest on the other. At the antinodes, where crests and troughs are in phase, water motion is predominantly up and down. The water motion is stopped when a crest or trough is at its peak, and horizontal motion is greatest when the surface of the water is perfectly flat. Arrows are included in Figure 3–17 to show relative water motion during successive stages in a standing wave.

Many of us experimented with the idea of a standing wave, but we did not put it in quite those terms. When you take your next Saturday night bath, fill the tub about halfway. Sit near one end of the tub with your elbows along the sides and grip the edges firmly with both hands. As you slowly slide along the bottom, you will set up a standing wave in the tub. As you slide back and forth in rhythm with the wave, you will begin to reinforce the standing wave and slosh the water over the end of the tub (Figure 3–18).

Wave refraction   Wave refraction plays an important role in the distribution of wave energy along a coast. If you fly over a coast, or look at an

FIGURE 3–18.
Sliding back and forth sets up a standing wave in a bath tub with wet results.
Drawing by C. W. L.

FIGURE 3–19.
A pencil in a glass of water and
light passing through a lens
demonstrate refraction of light
rays moving through water
and glass.
Drawing by C. W. L.

air photo taken of waves as they approach the shore, wave crests bend and become aligned parallel with the beach. The bending of wave crests as waves slow down is called refraction.

Refraction is common in many aspects of everyday life. If you place a pencil in a glass of water, the pencil appears to be bent at the surface of the water (Figure 3–19). Light rays travel more slowly in water than in air and are refracted when they leave the water, giving the illusion that the pencil is bent. As light rays pass through a lens, they are refracted so that the rays converge or diverge, depending on the type of lens (Figure 3–19). Camera and eye glass lenses provide examples of applications of the laws of refraction in the field of optics.

When a wave crest approaches the shore at an angle to the beach, the end of the crest that first enters shallow water starts to slow down and the wave crest is bent or refracted (Figure 3–20). A wave ray drawn at right angles to the wave crest points in the direction the wave is heading. In deep water, wave rays are straight and perpendicular to wave crests. As wave crests are refracted, wave rays spread apart, or diverge, along the beach (Figure 3–20).

When waves approach the shore, wave energy is constrained between two wave rays. Wave energy travels forward in the direction of wave rays and is not transmitted along the crest or across the rays. If two rays come closer together, or converge, as they approach the shore, wave energy is compressed into a shorter distance along the crest and wave height increases. If wave rays spread apart, or diverge, wave energy is stretched out along the crest and wave height decreases.

The effects of refraction are shown for a coast with a bay at the mouth of a river and a headland extending into the ocean (Figure 3–21). The bottom contour map shows the depths of water in the bay and off the headland. As the waves approach the coast, waves near the headland slow down and wrap around the point of land. The waves in the middle of the bay continue in deep water until they are near the head of the bay. Along the

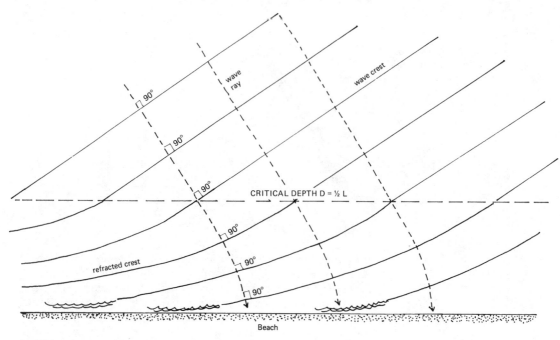

FIGURE 3–20. Waves are refracted in shallow water and wave crests are aligned more close to parallel with the shore. Wave rays at right angles to the crests bend toward the shore.
Drawing by C. W. L.

FIGURE 3–21. Waves are refracted along a coast with bays and headlands. The wave rays converge on the headland resulting in the erosion of cliffs. In the bay, wave rays diverge and spits are deposited.
Drawing by C. W. L.

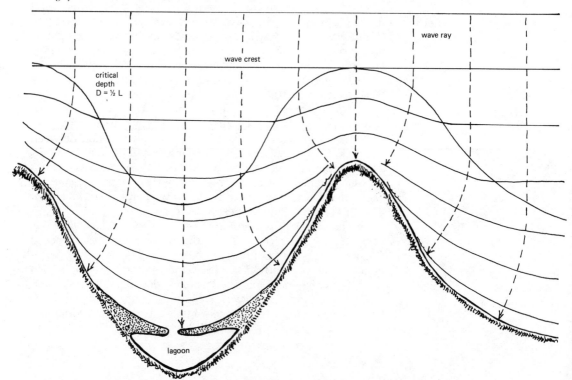

margins of the bay, waves are refracted as they enter shallow water.

In deep water, wave rays are a uniform distance apart and at right angles to the crests. Where the wave crests are refracted around the headland, the wave rays converge. The distance between rays decreases and wave height increases. Therefore, where wave energy is focused on the headland, wave height increases. The concentration of wave energy at the headlands results in coastal erosion and the formation of cliffs.

In the bay where the wave rays diverge, the distance between the rays increases. Wave energy is spread out in the bay where wave rays diverge and, therefore, the height decreases. Sand bars and spits are deposited in the bay, where the material is transported from the headland. Longshore currents sweep along the edge of the bay and carry sand and gravel from the headland to the spits and bars in the bay.

Wave diffraction

When waves sweep past a point of land or the end of a breakwater, a circular wave is generated that sets up an interference pattern called wave diffraction. Wave diffraction is of interest in harbor design, because it produces an increase in wave height at the entrance to the harbor and allows some wave energy to leak into the harbor behind the breakwater.

In order to understand wave diffraction, it is necessary to construct a wave diffraction diagram. A series of waves pass a breakwater that extends toward the center of the map (Figure 3–22). It is assumed that the water around the breakwater is relatively deep so that the effects of shoaling do not have to be considered. Seaward of the breakwater, waves are straight and parallel to the breakwater.

The wave diffraction pattern develops after waves have passed the tip of the breakwater. As a wave passes the breakwater, a circular wave is generated with a radius equal to the wave length of the incoming wave. The left half of the incoming wave is blocked by the breakwater, and some of the wave energy is transmitted into the circular wave.

A diffraction pattern is set up by the interaction between the incoming wave and the circular wave (Figure 3–22). The dotted line behind the breakwater is called the geometric shadow. The wave height along the geometric shadow is equal to one-half the height of the incoming wave ($H = \frac{1}{2} H_o$). The incoming waves are tangent to the circular waves along the geometric shadow. Behind the breakwater, the wave height drops off progressively from one-half H_o along the geometric shadow to zero behind the breakwater.

To the right of the geometric shadow, crests of the incoming and circular waves cross, forming a wave diffraction pattern. As the waves move forward, the crests of the incoming and circular waves cross along phase lines (Figure 3–22). The phase lines are plotted as dashed lines extending from the tip of the

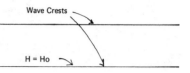

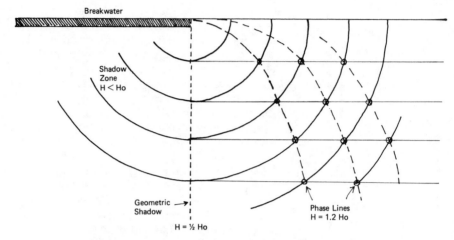

FIGURE 3–22. A wave diffraction pattern is produced when waves pass the tip of a breakwater. Circular waves intersect incoming waves along phase lines. Wave height drops to the left of the geometric shadow.
Drawing by C. W. L.

breakwater. Wave crests appear as humps, and troughs as hollows that move forward along the phase lines. The height of the circular wave is added to the incoming wave to give a maximum height of about 1.2 H_o along the phase line.

Midway between phase lines, the crests of the incoming waves are lined up with the troughs of the circular waves. Along this line, the waves are out of phase and the wave height is less than the incoming wave height. Although the wave diffraction pattern near a breakwater can be predicted by drawing the incoming and circular waves, it gives the harbor entrance a very choppy appearance.

When two breakwaters are brought together, the diffraction patterns of the two sides are superimposed (Figure 3–23). The phase lines cross in the entrance, giving the harbor a chaotic appearance. On the seaward side of each breakwater, the incoming waves are reflected and a standing wave is set up, giving increased wave heights. The distance between and angle of breakwaters determine the location and intersections of phase lines. By careful planning and design, it is possible to reduce the turbulence at the harbor entrance.

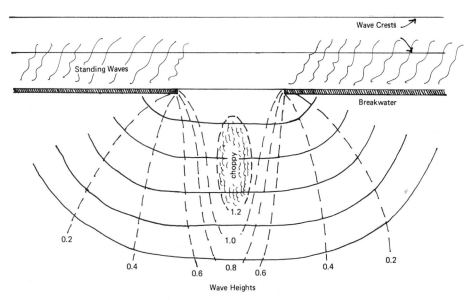

FIGURE 3–23. The combined diffraction patterns from two breakwaters produce a choppy area in the harbor mouth, with standing waves seaward of a breakwater.
Drawing by C. W. L.

Summary

Most waves on the surface of the ocean are generated by winds blowing across the water. Waves are described by several properties, including wave length, *L,* wave height, *H,* wave period, *T,* and phase velocity or speed of the wave, *C.* The shape of a wave is expressed as wave steepness, which is the ratio of the height to length *(H/L).*

Large waves in deep water are formed by storms in the open ocean. The wave spectrum, or range of wave lengths and heights produced by a storm, is determined by the wind speed, duration, and fetch. The wind speed depends on the size of the storm. Wind duration is the length of time the storm blows, and fetch is the distance the storm blows over water. The longer waves travel at a higher speed and move out of the storm area as swells.

Several changes take place when waves move from deep to shallow water. In shoaling waves, wave length and phase velocity decrease, and height first decreases, then increases as waves approach the beach. The waves break on the beach, producing plunging, spilling, collapsing, or surging breakers, depending on the bottom slope. Waves are reflected on a steep shore, forming standing waves. Wave crests are bent or refracted as they slow down in shallow water. When waves pass a point of land or the tip of a breakwater, a circular wave pattern is established that interacts with incoming waves, forming wave diffraction. Wave diffraction results in a choppy area within the mouth of a harbor.

Selected Readings

BASCON, W. *Waves and Beaches.* Garden City, N.Y.: Anchor Books, Doubleday, 1980. A clear introduction to waves and beaches for the general public. Emphasis on the Pacific coast.

GROSS, M. G. *Oceanography—A view of the Earth* (3rd ed.). Englewood Cliffs, N. J.: Prentice-Hall, Inc., 1982. A clear explanation of deep- and shallow-water waves along the coast.

KINSMAN, B. *Wind Waves—their Generation and Propagation on the Ocean Surface.* Englewood Cliffs, N. J.: Prentice-Hall, Inc., 1965. A mathematical treatment of water waves designed for advanced students. A clear introductory chapter on waves for the layperson.

NEUMAN. G. AND W. J. PIERSON, JR. *Principles of Physical Oceanography.* Englewood Cliffs, N. J.: Prentice-Hall, Inc., 1966. A mathematical treatment of wind waves and swells along the coast.

VAN DORN, W. G. *Oceanography and Seamanship.* New York: Dodd, Mead and Co., 1974. A practical introduction to wind waves and wave forecasting for sailors.

WIEGEL, R. L. *Oceanographical Engineering.* Englewood Cliffs, N. J.: Prentice-Hall, Inc., 1964. An introduction to shoaling, refraction, and diffraction with engineering applications.

World Meteorological Organization. *Handbook on Wave Analysis and Forecasting.* Geneva, Switzerland: Publication WMO No. 466, 1976. Practical methods and techniques for making wave forecasts.

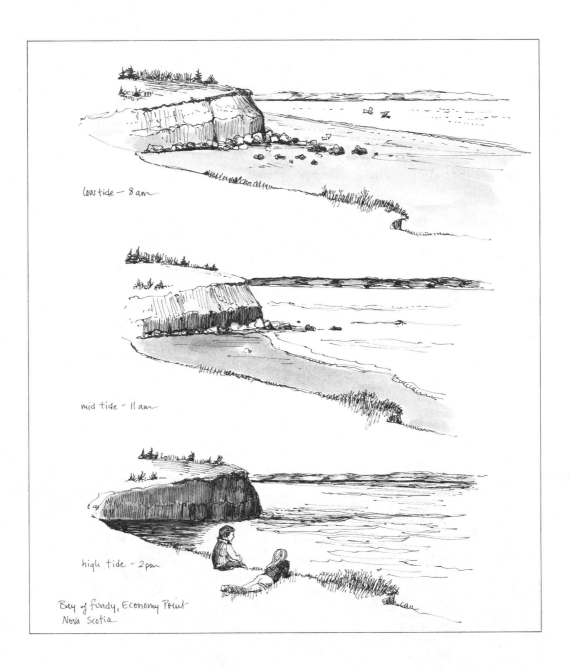

low tide – 8 am

mid tide – 11 am

high tide – 2 pm

Bay of Fundy, Economy Point
Nova Scotia

FOUR
Tides

Introduction to Tides

The daily rise and fall of the tides has an influence on many aspects of life along the coast. In Nova Scotia on the Bay of Fundy, fishing boats are left high and dry when the tide is out (Figure 4–1). At the mouth of the Columbia River on the Washington coast, large ships must wait for high tide in order to pass over the bar. The spawning and breeding of grunion on the California coast and horseshoe crabs on Delaware Bay also coincide with the high spring tides. Strong tidal currents must be taken into consideration for sailing and navigation in many coastal areas. An understanding of the forces and mechanisms producing the tides is essential for making reasonable tidal predictions.

Although tide tables and current charts are available for most coastal areas of North America and several other parts of the world, it is still a difficult problem to predict tides where accurate records are not available. The tide tables are calculated from the analysis of water level records from a tide gauge over a period of several years. When the records are processed on a computer using harmonic analysis, it is possible to examine past records to make predictions about time and height of future tides. Where tide records are not available, it is possible to make extrapolations from nearby tide recording stations to get a reasonable estimate of the tides. Storms and local changes in barometric pressure will change the tide pattern in shallow water, influencing the time of high and low water and the range of the tides.

FIGURE 4–1. At low tide, fishing boats are left stranded in the mud at Cape Blomidon on the Bay of Fundy, Nova Scotia. *Photo by W. T. Fox.*

The daily pattern of tides is familiar to anyone who has lived or worked along the coast. On the Atlantic coasts of North America and Europe, there are two high and two low tides each day, with an interval of 12 hours and 25 minutes between successive high tides. The twice daily tides, with about equal highs and lows, are known as semidiurnal tides. Along the coast of the Gulf of Mexico, there are one high and one low tide each day, with an interval of 24 hours and 50 minutes between successive high tides. The single daily tidal cycle is known as a diurnal tide.

The mixed tides, which are a combination of diurnal and semidiurnal tides, are well displayed along the Pacific coast of North America from Alaska to Mexico. The mixed tides have a diurnal inequality with a higher-high, lower-high, higher-low, and lower-low tide each day. Many organisms in the low inter-tidal zone are exposed only once each day, at the lower low tide. When large ships pass over the bar at the mouth of the Columbia River, it is often necessary to wait for the higher high tide to provide the necessary clearance.

Tidal range is the vertical difference in tide level between the high and low tides. Tidal amplitude is one-half the tidal range. Mean tide level corresponds to mean sea level, which is the average sea level without any tides. Mean high tide is the average high tide level and mean low tide is the average low tide level. Mean low tide is used as the base level for many coastal charts and for most tide tables. When ships move in and out of a harbor, mean low tide is a more critical depth than mean sea level.

Spring and neap tides refer to the differences in tidal range during the month. The greatest tidal ranges occur during spring tides, a few days after the new and full moons. The derivation of spring tides comes from the Saxon word, *springan,* which means a rising or welling of water and has nothing to do with spring season. Neap tides have the smallest tidal range and coincide with the first and third quarters of the moon.

Ebb and flood tides generally refer to the currents associated with tides. At a tidal inlet, flood currents flow into the inlet and ebb currents flow out to sea. At low tide, ebb currents are still flowing seaward in parts of the estuary. Low-tide slack or still water, when the current reverses, occurs a few minutes to an hour after low tide and lasts for 5 to 15 minutes. As water rises in the inlet, flood currents flow through the inlet mouth. Flooding continues through high tide with high-tide slack occurring a short time after high tide. The ebb current flows from the high-tide slack till the next low-tide slack.

Tidal forces

The forces that produce the tides in the ocean are generated by combined forces of the earth, moon, and sun. Two forces, gravity and centrifugal force, are responsible for the rise and fall of the tides. The water that falls as rain and snow on the continents flows to the oceans under the pull of the earth's gravity. The water fills in the low places in the ocean basins and forms a water layer that covers almost four-fifths of the earth's surface.

The sun and moon exert gravitational pulls on the water in the oceans. The forces of gravity are opposed by centrifugal forces, which are produced by the motions of the earth, moon, and sun. If the forces of gravity were unopposed, the moon would be pulled into the surface of the earth, and the earth would be pulled into the sun. The motions of the planetary bodies balance the force of gravity so that the moon remains in orbit around the earth, and the earth remains in orbit around the sun. The water in the oceans is free to flow and moves in response to the forces of the sun and moon.

Instead of considering the entire volume of water in the oceans, it is easier to think of the ocean as composed of small parcels of water, each free to move under the influence of the sun and moon. The parcels of water that are located at different positions on the earth's surface are also different distances from the center of the sun and the moon. The movements of the individual parcels of water are combined to give us a picture of the movement of the tides.

Force of gravity.   Newton's Law of Gravitation is used to explain the tidal forces on the surface of the earth. According to the Law of Gravitation, the force of attraction, *F,* between two bodies is directly proportional to the masses of the two bodies and inversely proportional to the square of the distance between them.

$$F = G \frac{M_1 M_2}{R^2}$$

where *G* is the gravitational constant, M_1 and M_2 are the masses of the two bodies, and *R* is the distance between the centers of mass of the bodies.

Gravity due to the earth's attraction always pulls toward the center of the earth. Due to the rotation of the earth about its axis, the shape of the earth has been slightly distorted by centrifugal force. Therefore, the earth is flattened at the poles and expanded around the equator, forming an oblate spheroid. Since the poles are closer to the center of the earth, the pull of gravity is slightly greater at the poles than at the equator. As you move from the equator toward the poles, there is a small but steady increase in gravity, but the gravity is constant around each circle of latitude.

The moon's gravity has an effect on the earth's tides, because each parcel of water is pulled directly toward the center of the moon (Figure 4–2). Since the moon's gravity is determined by square of the distance from the parcel of water to the

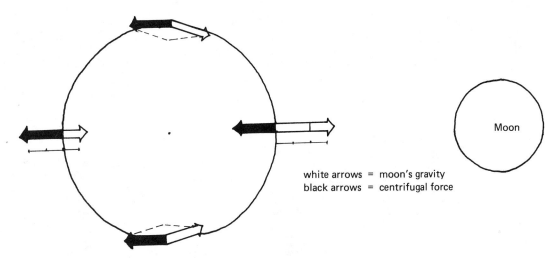

white arrows = moon's gravity
black arrows = centrifugal force

FIGURE 4–2. The tidal bulge on the earth's surface is produced by the moon's gravity and centrifugal force. On the side of the earth facing the moon, the moon's gravity is greater than centrifugal force, resulting in a tidal bulge. On the side of the earth away from the moon, the centrifugal force is greater than the moon's gravity, which also produces a tidal bulge on the opposite side of the earth from the moon. *Drawing by C. W. L.*

center of the moon, the water parcels closer to the moon will experience a stronger pull of gravity. The strongest pull of the moon's gravity occurs when the moon is directly overhead at the zenith *(Z)*, and the weakest moon gravity will be felt when the moon is on the opposite side of the earth at the nadir *(N)* (Figure 4–2). If you intersect the earth by a series of planes that are perpendicular to a line between the centers of the earth and the moon, the moon's gravity is constant along the circles formed by the intersections of the planes and the earth. Along each circle, the force of the moon's gravity is pointed toward the center of the moon.

Centrifugal force. Centrifugal force is a force directed outward from a center, which is exerted by a body moving in a curved path. For example, when you go around a curve at high speed in a car, your body is forced against the door by centrifugal force. When you attach a ball to the end of a string and swing it around your head, the ball is pulled outward by centrifugal force.

When thinking about the effect of centrifugal force on the earth and the moon, it is interesting to review what is known about manmade satellites launched into orbit around the earth. When a rocket places a satellite into orbit around the earth, the speed and mass of the satellite determine the radius of the orbit. If the speed of the satellite is decreased, the radius of the orbit decreases and eventually the satellite will spiral into the earth's atmosphere and burn up. If the speed of the satellite is increased, the radius of the orbit will enlarge, and if the speed is increased enough, the satellite will eventually escape the gravitational field of the earth.

The moon can be considered a large satellite in orbit around the earth. The mass of the earth is about 81.5 times the mass of the moon. Because the moon is such a large satellite, it also has a noticeable effect on the earth. Instead of the moon orbiting around a stationary earth, the earth and moon are orbiting around a common center. Because the earth has 81.5 times the mass of the moon, the moon is located 81.5 times as far from the center of rotation as the center of the earth (Figure 4–3). For example, if the earth and the moon were on opposite sides of a seesaw, the moon would have to be 81.5 times as far from the balance point as the earth.

The average distance between the center of the earth and the center of the moon is 384,400 kilometers (238,712 miles) (Figure 4–3). Therefore, the balance point, or center of rotation, between the earth and moon would be 4660 kilometers (2895 miles) from the center of the earth (384,400/82.5 = 4660). The

FIGURE *4–3*. The earth and the moon orbit around a common axis of rotation located within the earth. The earth has a larger mass and is closer to the axis of rotation, which corresponds to a balance point on a seesaw.
Drawing by C. W. L.

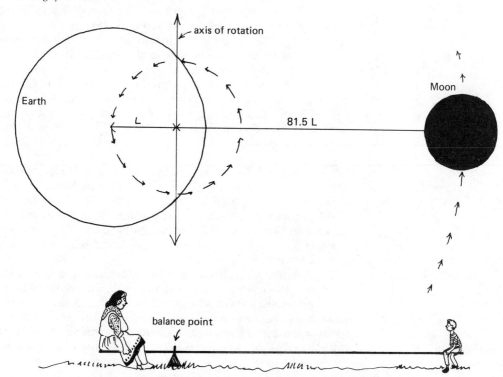

radius of the earth is 6376 kilometers, so the center of rotation would be located 1716 kilometers beneath the surface of the earth, or about a quarter of the way from the surface to the center of the earth (Figure 4–3).

It is easy to visualize the moon in orbit around the center of rotation because it appears to be orbiting around the center of the earth. Since center of rotation is located within the earth, the earth appears to wobble around the center of motion. As the earth moves around the center of rotation, the center does not remain fixed at one position within the earth. Each point on the surface of the earth moves in a circle with the same radius and experiences the same centrifugal force.

A cardboard cutout can be used to demonstrate the earth's wobble (Figure 4–4). To make the ring, fold a piece of card-

FIGURE 4–4. A cardboard ring is used to demonstrate the wobbling motion of the earth as it orbits around the axis of rotation. As the ring is moved around the circle, the arrow continues to point upwards. Each point on the outside of the circle (A and B) moves in a circle that has the same radius as the inner circle of the ring. Therefore, centrifugal force is the same everywhere on the surface of the earth.
Drawing by C. W. L.

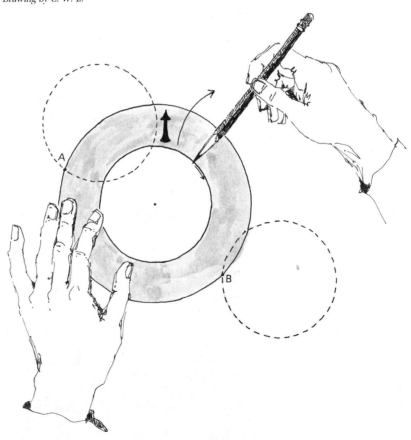

board, place the point of a drawing compass on the fold, and draw two semicircles, one with a radius of 2 inches and the other with a radius of 1 inch. Cut out the ring formed by the two circles. Draw an arrow pointing outward and point it up vertically. Place the ring on a table top, with the point of a pencil held firmly within the ring. Grip the ring with your left hand and slide the ring around the pencil so that the inner margin of the ring remains in contact with the pencil and the arrow you drew on the ring remains vertical (Figure 4–4). As you watch the ring, you will notice that any point on the outside ring also moves in a circle that has the same radius as the inner ring.

As the earth moves around the center of rotation, every point on the earth's surface moves in a circle that has a radius equal to the distance from the center of rotation to the center of the earth. Therefore, the centrifugal force is the same everywhere on the earth's surface, but the pull of the moon's gravity is stronger on the side of the earth facing the moon and weaker on the side away from the moon.

Astronomical effects Before studying the tidal effects of the earth, moon, and sun, it is a good idea to review the behavior of the three bodies. It is necessary to consider the orbit of the earth around the sun, the orbit of the moon about the earth, and the rotation of the earth about its axis. The orbital paths of the earth and moon are elliptical, which affects the pull of gravity on the tides. The axis of the earth is tilted, or inclined, with respect to the plane of the earth's orbit around the sun. The orbital plane of the moon is also tilted with respect to the earth's orbital plane. The tilt of the moon's orbital plane is known as the declination of the moon.

Tidal bulges—diurnal and semidiurnal tides. The difference between the pull of gravity and of centrifugal force produces tidal bulges on the surface of the earth. The pull of the moon's gravity is always directed toward the center of the moon and is greatest on the side of the earth facing the moon (Figure 4–2). The centrifugal force is equal everywhere on the earth's surface and is always away from the moon or opposed to the moon's gravity.

The water on the earth's surface is distorted into an egg shape, with bulges on the earth facing toward and away from the moon (Figure 4–5). The ocean surface has been depressed about 17.7 centimeters (7.0 inches) along the low-tide girdle and is uplifted about 35.4 centimeters (13.9 inches) in the high-tide bulges facing toward and away from the moon. The tides on the earth due to the moon would have a total range of 53.1 centimeters (1.7 feet), from a high of 35.4 centimeters to a low of −17.7 centimeters.

The high-tide bulges account for the semidiurnal tides with two high and two low tides in 24 hours and 50 minutes.

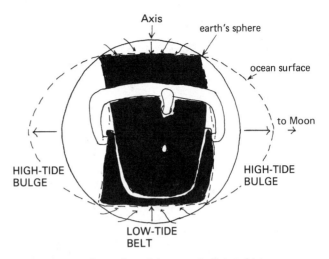

Axis

earth's sphere

ocean surface

to Moon

HIGH-TIDE
BULGE

HIGH-TIDE
BULGE

LOW-TIDE
BELT

FIGURE 4–5. The surface of the ocean is distorted into an egg
shape with the high-tide bulges facing toward and away from
the moon, and a low-tide belt encircling the earth.
Drawing by C. W. L.

As the moon proceeds in its orbit around the earth, the position
of the tidal bulges shifts each day to stay aligned with the moon.
The period of the semidiurnal tide is 12 hours and 25 minutes.
Therefore, the morning or afternoon high tide is 50 minutes
later the following day.

The side of the earth facing the moon is closer to the center of the moon and, therefore, feels a stronger pull of the
moon's gravity. The additional pull of the moon's gravity produces a diurnal tide with a period of 24 hours and 50 minutes.

Tidal bulges are also produced by the sun giving diurnal
and semidiurnal solar tides with periods of 24 and 12 hours.
The sun is 26.7 million times more massive than the moon, so
one would think that the sun would have a greater effect on the
earth's tides than the moon. However, tidal forces are proportional to the cube of the distance between the bodies. Since the
center of the sun is 388.6 times farther from the center of the
earth than is the moon, the effect of distance is 388.6^3, which
equals 58.6 million. The relative effects of the lunar and solar
tides are expressed in the following equation.

$$\frac{\text{Mass}}{\text{Distance cubed}} = \frac{26.7 \text{ million}}{58.6 \text{ million}} = .456 = 45.6\%$$

The size of the tidal bulge due to the sun will be 45.6 percent as
large as the tidal bulge due to the moon. The elevation of the
high tide due to the sun would be 16.1 centimeters
(6.3 inches), and the depression along the low tide would be
-8.1 centimeters (3.2 inches). The combined high tide when
the sun and moon are perfectly aligned with the earth would be

51.5 centimeters (20.3 inches), and the low tide would be −25.8 centimeters (10.2 inches). The maximum tidal range or difference between high and low tide would be 77.3 centimeters (about 2½ feet).

Phases of the moon—spring and neap tides. One of the major consequences of the motions of the earth, moon, and sun is the phenomenon of spring and neap tides. When the moon and sun are aligned with the earth, the high tides of the sun and moon are added together to produce spring high tides (Figure 4–6B). A new moon occurs when the moon and sun are on the same side of the earth. The sun is shining on the far side of the moon, leaving the dark side of the moon facing the earth. When the moon is on the far side of the earth from the sun, the sun's light is reflected off the surface of the moon toward the earth, giving a full moon. The highest spring tides occur when the sun and moon are on the same side of the earth at the new moon (Figure 4–6B). The spring tides occur about three days after the new and full moons, due to the inertial and frictional effects.

During the first and third quarters of the moon, lines connecting the center of the earth to the sun and moon form a right angle (Figure 4–6B). The high tides due to the sun and moon are out of phase, producing the minimum tidal range, which is

FIGURE 4–6A. Phases of the moon.

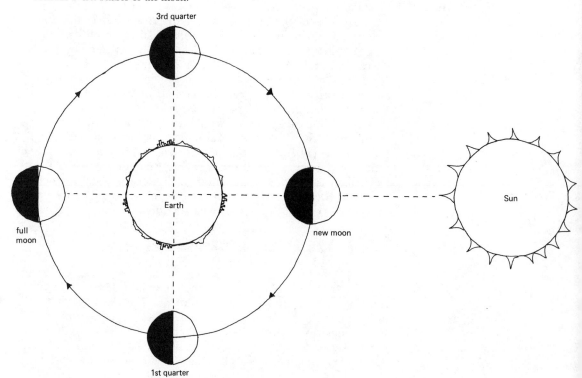

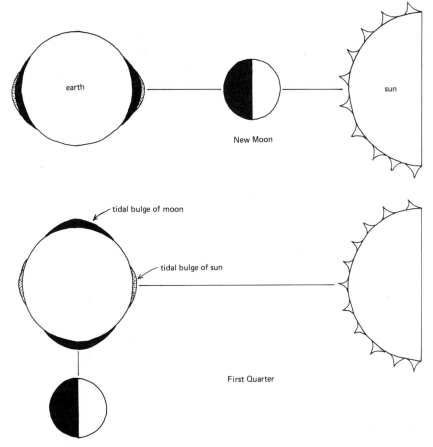

earth

New Moon

sun

tidal bulge of moon

tidal bulge of sun

First Quarter

FIGURE 4–6B. During the new moon (top) and full moon, the tidal bulges due to the sun and moon are superimposed, producing spring tides. During the first quarter (bottom) and third quarter of the moon, the lines connecting the center of the moon, earth, and sun form a right angle. The tidal bulges due to the sun and moon are out of phase, producing neap tides.
Drawing by C. W. L.

known as neap tides. In general, the spring tides are about 20 percent greater than the average tides, and the neap tides are about 20 percent less, but this varies considerably from one location to the next. The spring and neap tides occur twice each lunar month or about once every two weeks (Figure 4–7).

Inclination of the earth's axis—seasonal tides. The earth orbits around the sun in the ecliptic plane. The axis of the earth is inclined 23½ degrees from the ecliptic plane, which accounts for our progressive change of seasons. The tilt of the earth's axis remains the same in space as the earth revolves around the sun. In the Northern Hemisphere, during the summer solstice on

June 22, the sun is directly overhead at 23½° north latitude and produces warming in the Northern Hemisphere. At the winter solstice on December 22, the sun is directly overhead at 23½° south latitude. At the Vernal equinox on March 22, and the Autumnal equinox on September 23, the sun is directly overhead on the equator and the day and night are equal. The declination of the earth's orbit relative to the ecliptic plane has a seasonal effect on the tides. The spring tides are generally higher in the summer and lower in the winter.

Elliptical orbit of the earth—perihelion and aphelion. As the earth orbits around the sun, the orbital path is elliptical with the sun at one focus of the ellipse. The earth is nearest the sun at perihelion, which occurs in January. At aphelion, in July, the earth is at its greatest distance from the sun. At perihelion the earth is about 148.5 million kilometers from the sun, and at aphelion, the distance has increased to 152.2 million kilometers. At perihelion, the gravitational attraction of the sun is greatest and the solar component of the tides is at a maximum.

Elliptical orbit of the moon—perigee and apogee. The moon's orbit around the earth is also elliptical. At perigee, the moon is about 357,000 kilometers from the earth, and at apogee, it is about 407,000 kilometers. Although the departure from

FIGURE 4–7. Tidal ranges are plotted for spring and neap tides during the 28-day lunar month. Highest spring tides occur during the new moon, with lower spring tides at full moon. Neap tides occur at the first and third quarters of the moon.
Drawing by C. W. L.

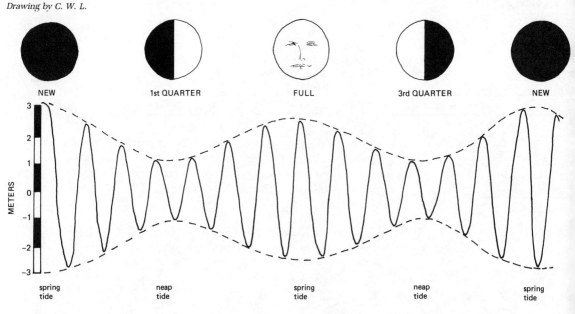

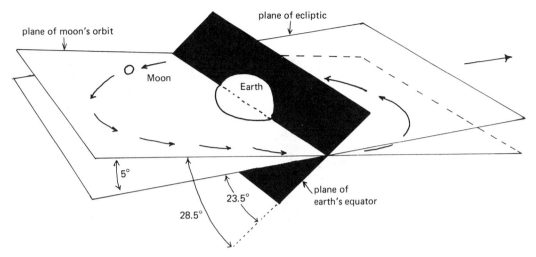

FIGURE 4–8. Highest spring tides occur once every 18.6 years, when the planes of the moon's orbit and the ecliptic plane are aligned to give the maximum lunar declination, 28.5°.
Drawing by C. W. L.

a true circle is not great, it does have an observable effect on the tides because of the gravitational force. The tidal period of this variation is 27.55 days.

Declination of the moon. The moon's orbit is inclined about 5 degrees to the ecliptic plane containing the orbit of the earth around the sun (Figure 4–8). The declination of the moon produces a long-term variation in the tides with a period of 18.6 years. The intersection of the plane of the moon's orbit with the ecliptic plane forms a line that slowly rotates around the earth, completing a revolution in 18.6 years. For 9.3 years, the moon's declination increases until it reaches a maximum of 28° (23.5 +5 = 28.5°). At this time, the center line of the moon's tide-generating force moves north and south by 57 degrees within a month. For the next 9.3 years, declination of the moon decreases to 18.5° (23.5 − 5 = 18.5°), and the north-south variation during the month is only 37°.

Eclipses of the sun are also controlled by the declination of the moon. When an eclipse occurs on the surface of the earth with the moon blocking out the sun, another eclipse will occur 18.6 years later along a line one-third of the way around the earth. Therefore, an eclipse of the sun occurs at about the same location after three cycles, or once every 55.8 years.

Tidal Theories

It has been known since the time of the early Greeks that the rise and fall of the tides are coordinated with the phases of the moon. Isaac Newton introduced the law of gravitation to explain effects that the pull of the moon had on the ocean. Newton proposed the equilibrium theory of tides, which used the law of gravitation to explain high-tide bulges on the sides of the earth facing and away from the moon. Tides are generated when the solid earth rotates beneath the tidal bulges in the ocean. A second theory, the dynamic theory of tides, added standing waves controlled by the size and shape of the ocean basin to account for local differences in tides. With the dynamic theory of tides, it is possible to make reasonable predictions of tidal amplitudes on the open ocean, but for practical tide predictions along a coast, it is still necessary to perform an harmonic analysis on recorded tidal data.

Equilibrium theory of tides

The equilibrium theory of tides was first proposed by Sir Isaac Newton in 1686, in his *Philosophiae naturalis principia mathematica*. Newton realized that he was treating the tides as a static problem, which was only a rough approximation of the real process. About fifty years later, the Paris Academy of Science held a contest to determine the best mathematical and physical explanation for the tides. Bernoulli, one of the contestants, essentially completed the equilibrium theory of tides.

According to the equilibrium theory, tidal bulges are produced by the combined effects of gravity and centrifugal force. High tides are produced when the earth rotates on its axis beneath a tidal bulge and low tide occurs when the earth is beneath the low tide girdle.

Assumptions for the equilibrium theory.

The equilibrium theory of tides has several assumptions that should be examined. First, in the original theory, it was assumed that there was only one tide-producing body, the moon. In later modifications of the theory the effects of the sun were included to account for the semidiurnal tides. Second, it was assumed that the equilibrium shape was produced by a balance between the centrifugal force and the moon's gravity. Third, it was assumed that the tidal bulges remained stationary while the earth rotated on its axis. Fourth, Newton assumed that the surface of the earth was covered with a uniformly deep and frictionless ocean. Finally, he assumed that tides would be equal along a line of latitude since the range of the tide is determined by the inclination of the earth's axis relative to the moon's orbital plane.

Equilibrium tides with no lunar declination.

The equilibrium theory of tides has been worked out for several different declinations between the earth's equatorial plane and the orbital plane of the moon. In the first case, assume that there is no tilt between the earth's equatorial plane and the orbit of the moon. The moon remains directly over the equator, giving the maximum high and low tides (Figure 4–9).

As the earth rotates on its axis, each point on the earth's surface follows a circle of latitude parallel to the equator. A

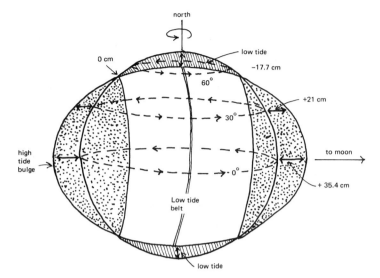

FIGURE 4–9.
The declination of the moon is zero when the moon's orbit lies in the same plane as the earth's equator. The high-tide bulges due to the equilibrium water surface remain on the equator when the lunar declination is zero.

point along the equator passes beneath two tidal bulges with each complete rotation of the earth. When the earth is beneath a tidal bulge, the high tide will be 35.4 centimeters above mean sea level. When the earth rotates a quarter turn on its axis, a point along the equator will pass beneath the low-tide girdle and the tide will drop to −17.7 centimeters. Therefore, an observer at the equator would encounter two high and two low tides each day, and the high tides would be about twice as high as the low tides (Figure 4–10A).

At 30° latitude, the high tide is lower, but the low tide remains the same (Figure 4–10B). According to the theory, the low tide will always be the same along the low-tide girdle. At 60° latitude, the tide is always less than zero but passes through the low-tide girdle twice each day (Figure 4–10C). The dividing line between the tidal bulge and depression occurs at about 52° latitude.

Equilibrium tides with maximum lunar declination. When the angle between the moon's orbit and the earth's axis is increased to the maximum possible, about 28°30′, a new pattern of tides emerges (Figure 4–11). On the equator, the earth passes beneath the tidal bulge, but not the highest part. Therefore, the highest tide is only 24.3 centimeters (Figure 4–12A). At 30° latitude, the earth starts near the maximum of the tidal bulge, passes beneath the low-tide girdle, then through a second high-tide that remains below mean sea level (Figure 4–12B). This gives a mixed tide with one high-high, and a low-high tide, which is characteristic of many locations around the Pacific Ocean. At 60° latitude, there is only one high and one low tide each day, but the duration of high is short and the low lasts for a longer time (Figure 4–12C).

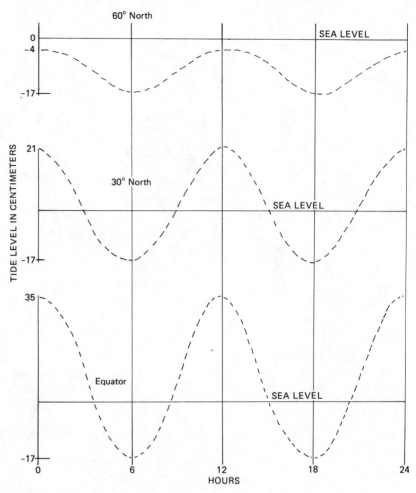

FIGURE 4–10. Plots of theoretical equilibrium tides with no lunar declination for 24 hours at three latitudes: (A) equator, (B) 30° north latitude, and (C) 60° north latitude.

FIGURE 4–11.
The equilibrium water surface is tilted 28½° north of the earth's equator at the time of the maximum declination of the moon.

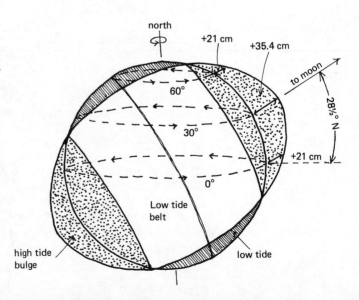

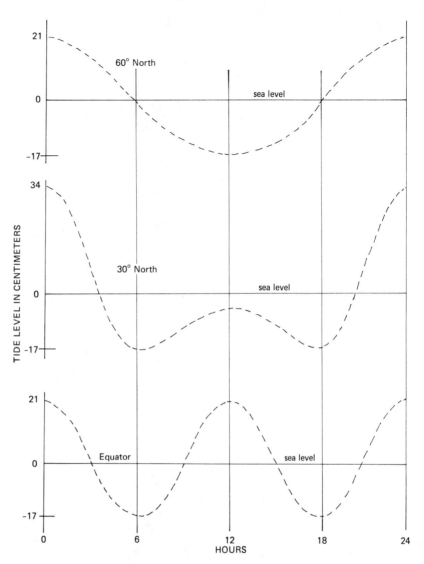

FIGURE 4–12. Plots of theoretical equilibrium tides with maximum lunar declination (28½°) at three latitudes: (A) equator, (B) 30° north latitude and (C) 60° north latitude.

The occurrence of semidiurnal tides at low latitudes, and the diurnal tides at high latitudes is in general agreement with actual tidal measurements, but it is masked by local effects of the tides. As the earth rotates about its axis, the moon is also revolving about the earth. Therefore, the interval between successive high tides is 12 hours and 25 minutes.

Although there are many shortcomings in the equilibrium theory of tides, it is studied as the first reasonable attempt to explain the tides. The principles gained from a study of Newton's theory provide mechanisms for generating a standing wave in the dynamic theory of tides.

Dynamic theory of tides   In the dynamic theory of tides, large standing waves in the ocean account for the tides along the coast. The dynamic theory was first proposed by Laplace, a French scientist and mathematican. The forces proposed for the equilibrium tidal theory provide a driving mechanism for dynamic tides. According to the equilibrium theory, the earth was covered with a deep layer of water. In the dynamic theory, the surface of the earth is divided into a group of relatively shallow ocean basins that are bounded by continents. Waves generated by the sun and the moon are trapped within the ocean basins and slosh back and forth setting up standing waves. In the equilibrium theory of tides, the tides move freely in the ocean without inertia. According to the dynamic theory of tides, inertia and the Coriolis effect are responsible for the complicated tidal behavior in the oceans.

The tidal waves produced by the moon's gravity and centrifugal force are not the tidal waves read about in the newspapers. Catastrophic waves that inundate a coastline have been called tidal waves, but are now referred to as tsunamis, from the Japanese word for harbor wave. The tsunamis are caused by landslides or motions on the ocean floor during large undersea earthquakes. During a severe hurricane or an extreme coastal storm, flooding along coastal lowlands caused by a storm surge is sometimes referred to as a tidal surge. In contrast, the more subtle wave due to tidal forces is a low-amplitude, long-period wave that produces high and low tides along coasts.

Standing waves in the ocean.   The waves produced by tidal forces have lengths that are equal to the width of the ocean. Therefore, the waves are considered shallow-water waves. The waves move from east to west as the earth rotates on its axis. When the wave is reflected off the western boundary of the ocean basin, it intercepts the next incoming wave and sets up a standing wave (see Chapter Three).

Every body of water has a natural period of oscillation for a standing wave. The period of oscillation is determined solely by the length and depth of the ocean basin. It is also possible to set up secondary standing waves that are harmonics of the fundamental period. For example, if the fundamental standing wave has a period of 12 hours, the second harmonic would have a period of 6 hours, and the third harmonic, a period of 4 hours. If the length of the ocean corresponds closely to one of the harmonics, a resonant condition will result, producing a forced standing wave with exceptionally high tides.

Consider a rectangular ocean in the Northern Hemisphere that is twice as long in the east-west direction as it is wide (Figure 4–13). If the natural period of oscillation for the basin is 12 hours, the standing wave will oscillate through one complete cycle in 12 hours. The initial water surface is high at the east and west ends of the basin, and low in the center. Nodal lines are present one-fourth and three-fourths of the distance from the west end of the basin (Figure 4–13).

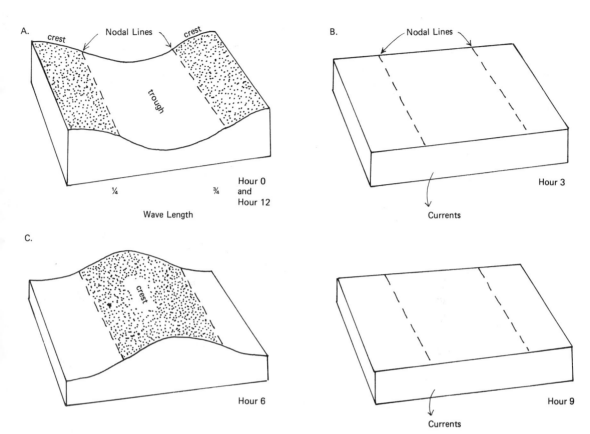

FIGURE 4–13. A standing wave with a period of 12 hours is generated in a rectangular basin. The width of the basin is equal to the wave length of the standing wave. Nodal lines are present at ¼ and ¾ wave lengths. The surface of the water and direction of currents are shown at 3-hour intervals. *Drawing by C. W. L.*

Amphidromic tides. Thus far, we have considered the water movement in an east-west ocean basin without taking into account the rotation of the earth. Due to the Coriolis effect, any moving object veers to the right in the Northern Hemisphere, and to the left in the Southern Hemisphere (See Chapter 2). The water that is converging toward the center of the basin at hour 3, or diverging toward the ends at hour 9, is forced to the right by the Coriolis effect (Figure 4–14). At hour 3, the water in the west half of the basin moves to the south, and the water in the east half of the basin moves to the north due to the Coriolis effect. If you follow the position of high tide in the basin, it moves counterclockwise (west, south, east, north, and west) (Figure 4–14).

The progression of high tides is plotted by placing the hour adjacent to the location of the high tide (Figure 4–15). The nodal point at the center is called an amphidromic point. Lines

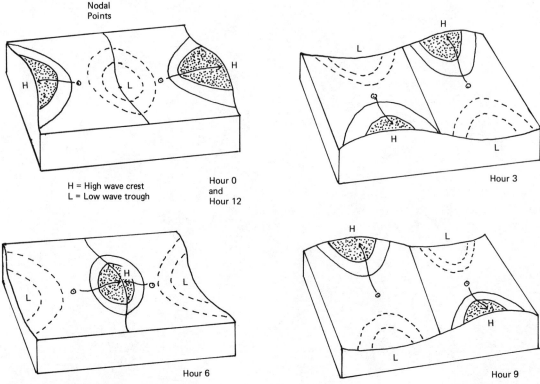

Nodal
Points

H = High wave crest
L = Low wave trough

Hour 0
and
Hour 12

Hour 3

Hour 6

Hour 9

FIGURE 4–14. The standing wave shown in Figure 4–13 is distorted by the Coriolis effect due to the rotation of the earth. The high tide rotates in a counterclockwise direction around the nodal point from hours 0 through 9.
Drawing by C. W. L.

FIGURE 4–15. Cotidal lines radiating from the nodal or amphidromic point indicate times of high tide. Corange lines forming concentric circles around the amphidromic point are lines of equal tidal range. The tidal range increases outward from the amphidromic point.
Drawing by C. W. L.

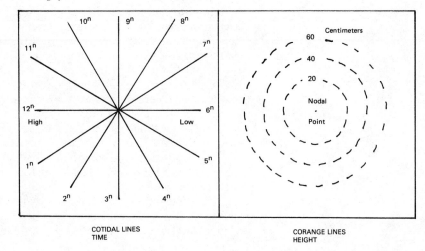

COTIDAL LINES
TIME

CORANGE LINES
HEIGHT

radiating from the amphidromic point are known as cotidal lines (Figure 4–15). It is high tide at all points along the cotidal line, and low tide along the cotidal line on the opposite side of the amphidromic point. For example, when it is high tide along the cotidal line for hour 3, it would be low tide along the cotidal line for hour 9.

The tidal range or difference between high and low tides is determined by the distance from the amphidromic points (Figure 4–15). At the amphidromic point, there is no vertical motion and the tidal range is zero. Away from the center along a cotidal line, the tidal range increases. When points of equal tidal range are connected, they form a series of concentric circles, which are called corange lines.

Corange and cotidal lines are plotted for the Atlantic and Pacific Oceans to show the location of the amphidromic points and the tidal ranges for the lunar semidiurnal tides (Figure 4–16). In the North Atlantic, an amphidromic point east of Newfoundland influences the east coast of United States. The width of the Pacific corresponds more closely to a wave length for a diurnal tide, with a second harmonic for the semidiurnal tide. Therefore, a mixed tide with diurnal and semidiurnal components about equal is present along the Pacific coast of North America. The length and depth of the Gulf of Mexico correspond to a diurnal tide, producing a single high and low tide each day.

FIGURE 4–16. Corange and cotidal lines around several amphidromic points in the world's oceans.

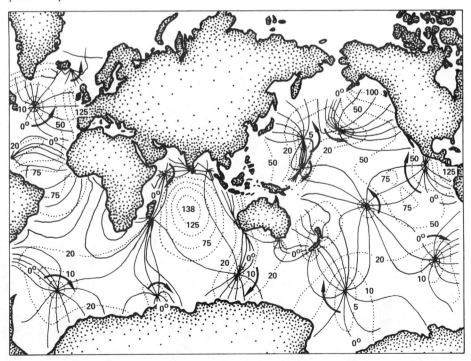

The dynamic or amphidromic tidal theory makes it possible to make approximate tide predictions on the open ocean for many parts of the world. The cotidal and corange lines give the approximate times of high and low tides and the tidal ranges where the local topography is not a factor. In harbors and estuaries, it is necessary to resort to harmonic analysis of the local tide records to make reasonable predictions of the tides.

The ranges of tides around the world can be roughly classified by their spring tidal ranges into three major groups: Microtidal—less than 2 meters, Mesotidal—2 to 4 meters, Macrotidal—greater than 4 meters. Most of the open coasts of the world are either microtidal or mesotidal, with the macrotidal category being reserved for unusual conditions in bays or estuaries.

Bay of Fundy tides. The highest tides in the world are recorded on the Bay of Fundy along the coast of Nova Scotia, Canada. The southwest end of the bay has a wide mouth that opens onto the North Atlantic. To the northeast the bay splits into two smaller bays, Chignecto Bay and the Minas Basin. The Bay of Fundy has a natural period of about 6 hours that resonates with the North Atlantic amphidromic tide producing a standing wave in the bay. The resonating wave and the constricted ends of the bay produce spring tides with a range of 15.6 meters (51.2 feet) at the head of the Minas Basin (Figure 4–17).

Harmonic analysis of tides

The practical problem of predicting tides at harbors and along a coastline involves harmonic or Fourier analysis. The observed tides can be separated into a series of sine curves or tidal components. The periods of the tidal components are based on astronomical behavior of the sun and the moon. Each of the tidal components has a period that can be correlated with the phases of the moon, the elliptical orbits of the earth and moon, the inclination of the earth's axis, or the declination of the moon. The periods of some of the major semidiurnal and diurnal components are listed in Table 4–1. The complete list of partial tides contains 390 components, but the seven major components account for a large proportion of the tidal variability. Although the period is constant for each tidal component, the phase and amplitude describe a unique sine curve for each coastal locality (Table 4–1).

Semidiurnal components. The first four tidal components M_2, S_2, N_2, and K_2, are the semidiurnal components with periods near 12.00 hours (Table 4–1). The principal lunar component, M_2, has a period of 12.42 hours (12 hours and 25 minutes). The M_2 component gives us the time of high water when the moon passes directly overhead or on the opposite side of the earth. On each successive day, the first high tide due to the M_2 component arrives 50 minutes later. The principal solar component, S_2, marks the time when the sun passes directly overhead or on the opposite side of the earth and has a period of 12.00 hours. The

FIGURE 4–17. Broad tide flats, which are exposed at low tide
(8:00 am), are covered at high tide (2:00 pm) at Economy
Point on the Bay of Fundy, where the spring tide range is
about 15 meters (50 feet).
Photo by W. T. Fox.

Table 4-1. PERIOD, PHASE, AND AMPLITUDE FOR THE MAJOR TIDAL COMPONENTS AT THREE LOCATIONS DURING MARCH, 1936.

COMPONENT	PERIOD (IN HOURS)	IMMINGHAM, ENGLAND PHASE (IN DEGREES)	AMPLITUDE (IN CM)	DO SAN, VIETNAM PHASE (IN DEGREES)	AMPLITUDE (IN CM)	SAN FRANCISCO, CALIFORNIA PHASE (IN DEGREES)	AMPLITUDE (IN CM)
semi-diurnal							
M_2-Lunar	12.42	161	223.2	113	4.4	330	54.2
S_2-Solar	12.00	210	72.8	140	3.0	334	12.3
N_2-Lunar elliptic	12.66	141	44.9	99	0.8	303	11.5
K_2-Declination	11.97	212	18.3	140	1.0	328	3.7
diurnal							
K_1-Declination	23.93	279	14.6	91	72.0	106	37.0
O_1-Lunar	25.82	120	16.4	35	70.0	89	23.0
P_1-Solar	24.07	257	6.4	91	24.0	104	11.5

M_2 and S_2 components are in phase at full moon and new moon, producing the spring tides with the largest tidal range. The M_2 and S_2 components are out of phase at the first and third quarters when the tides have their lowest range, giving neap tides.

The N_2 tidal component, with a period of 12.66 hours, is based on the elliptical orbit of the moon. When the moon is nearest the earth, the M_2 and N_2 components are in phase, giving the highest tidal range of the month. The K_2 component accounts for the changes in the declination of the sun and the moon throughout their orbital cycle. The K_2 component has a period of 11.97 hours, quite close to the M_2 component. Therefore, it is a long interval between times when the M_2 and K_2 components are in phase.

Diurnal components. The three major diurnal components K_1, O_1, and P_1 operate on a daily cycle with periods close to 24 hours. The solar-lunar component K_1 accounts for the diurnal effect of the earth's declination. The K_1 component usually has the largest amplitude for the diurnal tides. The principal lunar diurnal component O_1 shows the effect of the greater tidal range on the side of the earth nearest the moon. The principal solar diurnal component, P_1, produces a diurnal effect with increased tides on the side of the earth nearest the sun. As with the semidiurnal components, the phase and amplitude of the diurnal components are influenced by the latitude and longitude of the station, but most of all by the shape and size of the ocean basin. The period is fixed for each tidal component, but the phase and amplitude vary with the location.

The phase for each tidal component is expressed in degrees from 0 to 360, showing the relative starting position within a tidal cycle for each tidal component (Table 4–1). The major semidiurnal components, M_2 and S_2, are in phase once every 13.66 days, producing the semidiurnal spring tides.

The amplitude for each tidal component is equal to the vertical rise or fall of water level above or below mean sea level (Table 4–1). When the semidiurnal tide components, M_2, S_2, N_2, and K_2 have large amplitudes, the typical semidiurnal tide is produced. However, if the S_2 is almost equal to M_2, the difference in range between the spring and neap tides is large. When the diurnal components K_1, O_1, and P_1 are large, a diurnal tide results. If both semidiurnal and diurnal amplitudes are large, a mixed tide is produced with large inequalities in the height and time of successive high and low tides.

Three examples are given to show how the tidal components are combined to give different types of tides. The predicted tides are plotted for Immingham, England; Do San, Vietnam; and San Francisco, California (Figures 4–18, 4–19 and 4–20). The three records are taken for March, 1936, a month that includes the spring equinox, when the highest spring tides of the year occur. The stages of the moon are plotted with full moon on March 9, and new moon on March 23.

Semidiurnal tides—Immingham, England. The typical semidiurnal tide is shown for Immingham, England (Figure 4–18), where high tides occur every 12 hours and 25 minutes, corresponding to the period of the principal lunar component, M_2. On March 12 and 26, the M_2 and S_2 components are in phase, producing spring tides with a range of 7.3 meters on March 26. On March 5 and 20, the M_2 and S_2 components are out of phase, giving rise to a neap tide with a range of about 2.5 meters. The effects of the diurnal components K_1 and O_1 can be seen with the daily succession or beat of the tides.

Diurnal tides—Do San, Vietnam. The tide prediction for Do San during March, 1936, is a good example of a diurnal tide (Figure 4–19). Do San is located in the Gulf of Tonkin, where the semidiurnal components have a small amplitude and do not have much influence on the tides. The two major diurnal components, K_1 and O_1, have relatively large amplitudes. The spring tides occur on March 11 and 25, when K_1 and O_1 are in phase, and reach a maximum on March 25, when the P_1 component is also in phase. The maximum tidal range at Do San is about 3.2 meters. Since K_1 and O_1 components have almost the same amplitude (Table 4–1), the difference in tidal range between the spring and neap tides is large.

FIGURE 4–18. Large semidiurnal tidal components, M_2, S_2, N_2, and K_2 result in semidiurnal tides at Immingham, England.

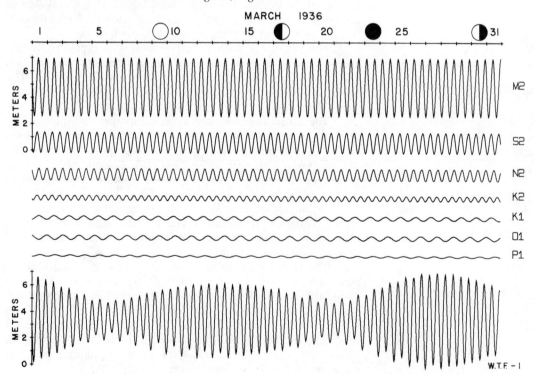

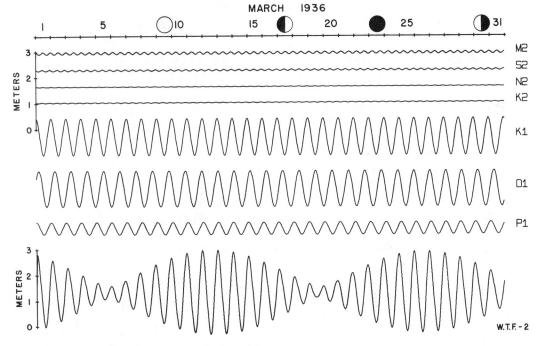

FIGURE 4–19. Large diurnal components, K_1, O_1, and P_1
produce diurnal tides at Do San, Vietnam.

Mixed tides—San Francisco, California. San Francisco
has a mixed tide with a larger semidiurnal than diurnal component (Figure 4–20). The spring tides following the full and
new moons, and the neap tides following the first and third
quarters are similar to the semidiurnal tides at Immingham.
The successive high or low tides can be quite unequal in amplitude. Each day has a higher-high tide, a lower-high tide, a
higher-low tide, and a lower-low tide. If you plan to study the
exposed tidal flats at spring tide in California, Oregon, or Washington, it is important to go out at the lower-low tide when the
flats are best exposed. At higher-low tide on the same day, the
flats will be covered by a meter of water.

The time of high tide along the coast may differ by an
hour or more from the predicted time printed in the tide tables.
With strong onshore winds, excess water can be piled into a bay
or lagoon. Therefore, the time of high tide would be about an
hour later. As the water drains out of the bay, the time of low
tide would also be delayed. The height of the observed tide can
also be higher than the predicted tide when water is piled up in
the bay by onshore winds. When you are planning a trip that is
contingent upon the tides, add about an hour to the predicted
time for high or low tide when strong winds are blowing
onshore.

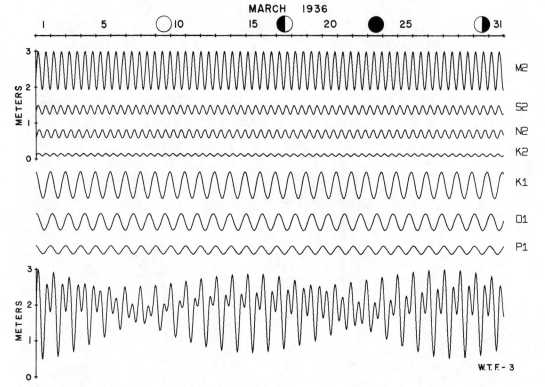

FIGURE 4–20. Large semidiurnal and diurnal components combine to form mixed tides at San Francisco, California.

Tidal Currents

As the tide level rises and falls, tidal currents are generated in the open ocean and along the coast. In the open ocean, the currents change direction with the rotating tidal wave (Figure 4–14). If no wind or external currents are present, a floating object will swing in a large counterclockwise oval, returning to its original position at the end of the tidal cycle. Along a coast, the rising tide floods the lowlands, generating the flood current. During the falling tide, the water drains from the lagoons and salt marshes, forming the ebb current.

Currents in a tidal inlet

The timing and speed of the tidal currents are closely related to the rise and fall of the tides (Figure 4–21). Starting with the tidal cycle at low tide, the flood current accelerates, reaching its maximum speed about midway through the flood cycle. The ebb current also accelerates during the falling tide, with the greatest speed at about mid-ebb. The dominant ebb and flood currents flow in separate channels formed by sand bars and shoals in the tidal inlet. An ebb or flood current can be dominant if it has either a higher speed or a longer duration of flow.

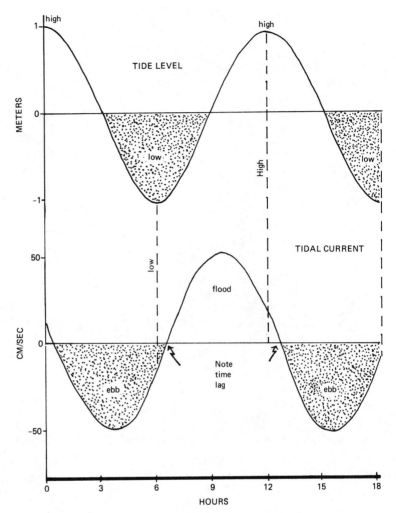

FIGURE 4–21. Tide level (top) is closely related to ebb and flood currents (bottom) in a tidal inlet. The reversal of current from ebb to flood usually occurs about ½ hour after low tide.
Drawing by C. W. L.

The development of flood and ebb channels leads to time-velocity asymmetry. Time asymmetry means that the maximum velocity does not occur at mid-tide, but occurs late in the tidal cycle. For example, the maximum flood velocity in the marginal flood channel often occurs an hour or more after mid-flood tide. In the main ebb channel, the ebb current flows down the middle of the inlet and the maximum ebb current often occurs late in the ebb cycle. The velocity asymmetry is due to the funneling effects of shoals and sand bars in the mouth of the inlet.

The strongest tidal currents in the world occur at the Saltstraumen maelstrom near Bodø in Norway's land of the midnight sun (Figure 4–22). The current runs through a narrow

FIGURE 4–22. The strongest tidal currents in the world occur
at Saltstraumen maelstrom near Bodø in Norway's land of
the Midnight Sun.
Courtesy of Norwegian National Tourist Office.

sound that separates two large fjords, forcing huge masses of
water in and out two times each day. The pressure is so strong
that the ground vibrates. The maelstrom provides excellent and
exciting fishing.

**Tidal currents in
the open ocean**

The tides in the open ocean are not restricted by nearby land
masses. Therefore, a circular current pattern is evident with the
current direction making an oval during each tidal interval. The
speed and direction of tidal currents at Nantucket Shoals Light
Ship are plotted in Figure 4–23. The speed and direction of the
current are plotted at 1-hour intervals, before and after high
and low tide. The tips of the arrows are connected in an oval
shape in a diagram that is called a hodograph (Figure 4–23).
The direction of the current vectors changes in a clockwise manner through the tidal cycle. If a log is anchored with an elastic

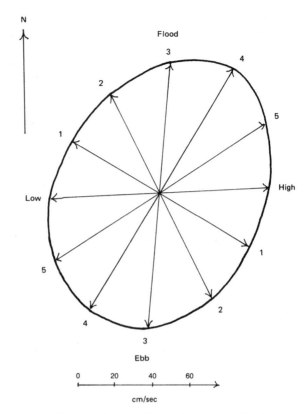

FIGURE 4–23. A hodograph is a plot showing direction of
tidal currents during a tidal cycle at Nantucket Shoals
Light Ship.
Drawing by C. W. L.

tether near the light ship, it will move in an oval, as shown by
the hodograph. If there are no additional currents or wind, the
log will return to its original position after the end of the tidal
cycle.

Summary The daily rise and fall of the tides are produced by the pull of
the sun and moon on the oceans. The moon's gravity pulls water
toward the center of the moon and is strongest on the side of
the earth facing the moon. The earth and moon rotate around
a common axis located within the earth, producing a centrifugal
force that is equal everywhere on the surface of the earth. The
effects of gravity and centrifugal force produce two tidal bulges
in the ocean, one facing and one away from the moon.

The equilibrium theory of tides assumes a uniformly deep
and frictionless ocean. The tides are produced when the earth
passes beneath the tidal bulges. A mixed tide results from the
declination of the moon's orbital plane relative to the earth.

The dynamic theory of tides assumes that large standing waves are formed in the oceans. Due to the Coriolis effect, the standing waves rotate around a nodal or amphidromic point in a counterclockwise direction in the Northern Hemisphere. The time and range of the tides are plotted as cotidal and corange lines on maps of the oceans.

Harmonic analysis is used to predict the time and range of tides along a coast. Semidiurnal and diurnal tidal components are added together to compute the tidal ranges.

Tidal currents are produced by rising and falling tides. Ebb and flood currents flow out of and into tidal inlets and estuaries. Tidal currents trace a circular or oval pattern in the open ocean.

Selected Readings

ANIKOUCHINE, W. A. AND R. W. STERNBERG. *The World Ocean—An Introduction to Oceanography* (2nd ed.). Englewood Cliffs, N.J.: Prentice-Hall, Inc., 1981. A nonmathematical explanation of the tides.

CLANCY, E. P. *The Tides, Pulses of the Earth.* Garden City, N.Y.: Doubleday and Company, 1968. A general book on all aspects of tides.

DEFANT, A. *Physical Oceanography, Volume II.* Oxford, England: Pergamon Press, 1960. An advanced text with several chapters devoted to analysis of tides.

GODIN, G. *The Analysis of Tides.* Toronto, Canada: The University of Toronto Press, 1972. A mathematical treatment of tides for advanced students in physical oceanography.

IPPEN, A. T. *Estuary and Coastline Hydrodynamics.* New York: McGraw-Hill Book Company, 1966. A good explanation of the Equilibrium Theory of Tides.

MACMILLAN, D. H. *Tides.* New York: American Elsevier, 1968. A clear explanation of tidal theories, with some basic mathematics, but not too deep.

NEUMANN, G. AND W. J. PIERSON, JR. *Principles of Physical Oceanography.* Englewood Cliffs, N.J.: Prentice-Hall, Inc., 1966. A mathematical explanation of tides.

TRICKER, R. A. R. *Bores, Breakers, Waves and Wakes, An Introduction to the Study of Waves on Water.* New York: American Elsevier, 1965. A nonmathematical introduction to tides with a good explanation of tidal bores.

Cape Cod Light, Truro, Massachusetts

FIVE
Beaches

Introduction to Beaches

A beach is a broad ribbon of sand that lies on the land along the edge of the sea, but in reality belongs to the ocean. The sand or gravel that forms the beach was derived from the erosion of sea cliffs, or carried to the sea by rivers, but the shape and form of the beach was molded by the sea. Waves and longshore currents pile the sand along the shore and shape it into beaches and longshore bars.

Man has turned to the beach as a break from the hectic pressures of modern life. The beach provides a restful setting for picnics, walks along the shore, and sun bathing, and also encourages more vigorous sports including swimming, jogging, surf casting, and surfboard riding. The cool sea breezes and nearby water make the beach an attractive place to enjoy a seaside vacation.

The beach plays an important role in the balance of nature. The beach provides a buffer zone that protects the coast from the onslaught of the waves. When storm waves break along the coast, the beach absorbs the wave energy. The sand is constantly shifted by the waves, forming new shapes on the beaches or nearshore bars along the coast. When sand is removed from the beach, the rocks along the coast are ground up to produce more sand. The sand on the beach is also moved along the shore by longshore currents. Some consider the beach a river of sand that flows along the coast.

Beach Materials

Most beaches are covered with gravel or sand that was derived from the erosion of nearby sea cliffs. The pebbles and sand grains on beaches are composed of a wide variety of rock types and minerals. The pounding and grinding action of waves has broken the rocks to produce a broad range of particle sizes and shapes. The grain sizes found on beaches range from huge boulders, more than a meter in diameter, to fine sand, a fraction of a millimeter across. The shapes range from rough, angular blocks, which were broken off sea cliffs, to smooth, polished pebbles, which are tumbled by the waves in the surf zone.

Particle size

The size of the individual rocks or sand grains is one of the most distinctive characteristics of a beach. A gravel or shingle beach consists of water-worn stones or pebbles that lie in loose sheets or beds on the seashore. A sand beach is covered with a fine debris of ground-up rocks consisting of small, loose grains, usually quartz. Broad tidal flats are often covered with fine sand or mud.

Gravel. The stones and rocks on a gravel beach are grouped into boulders, cobbles, pebbles, and granules (Table 5–1). Boulders are the largest rocks on a beach and include those with a diameter greater than 25 centimeters, about the size of a basketball. Granite boulders on Cape Ann, Massachusetts, were originally rectangular quarry blocks that were rounded by the waves (Figure 5–1). Cobbles are somewhat smaller, about the size of a grapefruit. Pebbles are about the size of ping pong or golf balls. The smallest grains of gravel, called granules, are about the size of small peas.

Table 5–1. GRAIN SIZES FOR PARTICLES, UNCONSOLIDATED SEDIMENTS, AND SEDIMENTARY ROCKS FOUND ON BEACHES.

SIZE	*PARTICLES*	*UNCONSOLIDATED SEDIMENTS*	*SEDIMENTARY ROCKS*
256 mm	Boulder		
	Cobble		
64 mm		Gravel	Conglomerate
	Pebble		
4 mm			
	Granule		
2 mm			
	Coarse Sand		
½ mm			
	Medium Sand	Sand	Sandstone
¼ mm			
	Fine Sand		
¹⁄₁₆ mm			
	Silt		
¹⁄₂₅₆ mm		Mud	Shale
	Clay		

FIGURE 5–1. Worn and rounded blocks of granite form a boulder beach at Cape Ann, Massachusetts.
Photo by W. T. Fox.

Sand. Most beaches are covered with sand ranging from $\frac{1}{16}$ to 2 millimeters in diameter. Coarse sand usually piles up into narrow, moderately steep beaches with fine sand forming broad, flat beaches. Medium sand is about the same size as table salt or granulated sugar.

Mud. Mud that covers the wide tidal flats is a combination of silt and clay. Silt consists of dust-size particles that resemble flour. Much of the silt found along the coast was formed by the grinding action of glaciers, but some was also formed by the chemical breakdown of igneous rocks in warm climates. Clay is the finest particle size, less than $\frac{1}{256}$ millimeter in diameter. It is impossible to distinguish individual grains of clay with the naked eye. Clay actually consists of tiny, flat flakes that resemble mica. Fresh clay is usually a light gray color and has a greasy feel, but it may be red if stained with iron, or black if it contains decayed organic matter.

Particle composition

A gravel beach is usually composed of rock fragments eroded from the adjacent sea cliff. Therefore, the gravel contains either igneous, metamorphic, or sedimentary rocks, depending on the types of rocks in the cliffs. A brief review of the basic rock types may be helpful for the identification of boulders, pebbles, or cobbles along the beach.

Igneous rocks. Igneous rocks (from *ignis,* the Latin word for fire) are formed by the congealing of magma, or melted rock, which has risen from deep within the earth. When magma has erupted on the earth's surface as lava or exploded into the air as ash, it forms volcanic igneous rocks. On continents, explosive volcanoes form light-colored volcanic rocks known as andesite, which are rich in silica. Volcanoes that erupt in the middle of the ocean form basalt, which is a dark green to black, fine-grained, volcanic rock, rich in iron and magnesium.

Plutonic igneous rocks are formed from magma that crystallized slowly beneath the earth's surface. Granite can be considered a coarse-grained equivalent to andesite, which contains crystals of quartz, feldspar, and mica. Granite and related plutonic rocks generally form within the roots of mountain ranges along the margins of continents. Gabbro is a coarse-grained basalt that cooled slowly within thick layers known as dikes or sills.

Sedimentary rocks. Sedimentary rocks are formed at the earth's surface, either by accumulation and later cementation of fragments of rocks, minerals, and organisms; or by precipitation and organic growth from seawater and other solutions. The most common sedimentary rocks include conglomerate, sandstone, limestone, and shale. A conglomerate is essentially a gravel that has been welded together to form a solid rock (Table 5–1). The pebbles in a conglomerate are cemented by silica or calcite, which was dissolved in water circulating within the pore spaces of the sediment. When grains of loose sand are cemented together, they are transformed into a sandstone. Shells of organisms that were living in the sea are cemented together to form limestone. Mud that was buried and compacted hardens to form shale. Sedimentary rocks laid down on the ocean floor are characterized by layers, or strata, and often contain fossils or imprints of shells.

Metamorphic rocks. Metamorphic rocks are formed when igneous or sedimentary rocks are subjected to intense heat and pressure deep within the earth. When igneous rocks are metamorphosed, they recrystallize, forming somewhat larger grains, but they are difficult to distinguish from the original igneous rocks. However, when shale is metamorphosed, it forms thin layers called foliation. The foliation consists of relatively coarse bands a few millimeters thick in a gneiss, or layers thinner than a sheet of paper in slate. Schist has foliation of intermediate thickness, between gneiss and slate. Foliated metamorphic rocks split along the foliation planes, forming thin slabs of rock. When sandstone is metamorphosed, it forms quartzite, a hard, massive rock. Limestone is recrystallized during metamorphism to form marble.

Sand grains. When coarse-grained rocks are broken down by the waves, the individual crystals are separated to form sand grains. Granite breaks down into quartz and feldspar, which are abundant on many sand beaches. Quartz is a colorless mineral that is hard and resistant to weathering. Most of the quartz grains on the beaches have been rounded and polished by the swash action of waves. Feldspar is a light-colored mineral that is not as hard as quartz and breaks down into small cleavage fragments or regularly shaped pieces. There are often dark specks in the sand, which are tiny fragments of iron minerals such as magnetite or ilmenite. Small grains of red or purple garnet are also common on many beaches and give the sand a reddish tint. When storm waves erode a sand cliff or dune, a large deposit of magnetite or garnet is often laid down along the base of the cliff (Figure 5–11).

On some volcanic islands, black or green sands are found along the beaches. The black sands were derived from very fine grained basalt that was extruded from a volcano. Some lavas contain the green mineral, olivine, which is common on some beaches in the Pacific.

White sand beaches in the tropics are covered with broken shell fragments or pieces of coral. The shells are composed of the mineral calcite (calcium carbonate), which was secreted by clams or snails. Calcite is a fairly soft mineral that rapidly breaks down into sand-size grains. However, when the individual sand grains composed of calcite are inspected with the aid of a microscope, it is still possible to see the texture of the shells.

On some tropical beaches, calcite grains are cemented together by calcium carbonate to form a resistant layer known as beachrock.

Pebble shapes   When rocks are exposed in cliffs along the coast, fragments are broken off by waves and piled up on the beach. The freshly broken rocks have rough edges and sharp corners, which are rapidly worn down by the tumbling action of the waves. Small fragments, which are broken-off boulders and cobbles, are ground down to pebbles and sand. As long as rocks remain in the surf zone, the average grain size is decreased as the rocks are worn, rounded, and polished.

Round pebbles.   Igneous rocks including granite and basalt produce spherical or egg-shaped pebbles. The granite pebbles are pink to light gray with a speckled appearance from the mixture of quartz, feldspar, and biotite (Figure 5–2). Basalt pebbles are very fine grained, and dark green to black, with smooth, highly polished surfaces. Fishing floats are often found among the pebbles.

Oval or flat pebbles.   Sedimentary rock pebbles, including sandstone, limestone, and shale, are generally smooth with oval shapes. Sedimentary rocks break into thin layers that result in flat pebbles. Sandstone pebbles are usually a medium to dark brown and often show color bands or coarse layers. Limestone pebbles are light to dark gray and often contain abundant fossil fragments (Figure 5–2). Shale pebbles come in a wide variety of colors, from brown to red, green, or black. Shale splits into thin layers, forming smooth disk-shaped pebbles that are excellent skipping stones.

Irregular pebbles.   Metamorphic rocks form more irregularly shaped pebbles than igneous or sedimentary rocks. Schist and phyllite break along foliation planes, producing rough pebbles (Figure 5–2). Pebbles from phyllite, schist, and gneiss are usually medium to dark gray and contain thin quartz veins. When the metamorphic rocks are heated and folded deep within the earth, the quartz flows and collects in thick veins or knots, which split out, forming quartz pebbles with a milky appearance. Quartzite and marble are light-colored rocks with a uniform texture that produces spherical or oval shapes resembling igneous pebbles.

Coastal Zones   A typical profile across a coastal area includes four major zones, starting on the land and heading toward the sea: (1) the coast marked by sea cliffs, dunes, or permanent vegetation; (2) the shore or beach; (3) the inshore or surf zone; and (4) the offshore zone or continental shelf (Figure 5–3). The character and

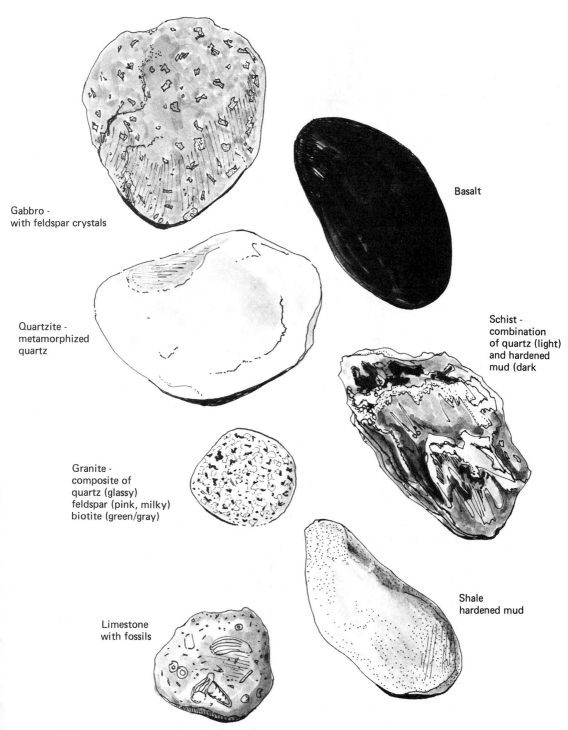

Gabbro -
with feldspar crystals

Basalt

Quartzite -
metamorphized
quartz

Schist -
combination
of quartz (light)
and hardened
mud (dark

Granite -
composite of
quartz (glassy)
feldspar (pink, milky)
biotite (green/gray)

Shale
hardened mud

Limestone
with fossils

FIGURE 5–2. A collection of pebble types common along the
shore includes gabbro, basalt, schist, shale, limestone, granite,
and quartzite (clockwise from the top).
Drawing by C. W. L.

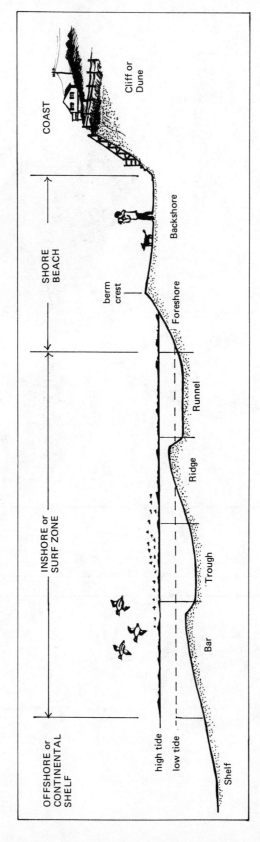

FIGURE 5-3. Profile across the coast.

width of the coastal zones vary with the type of coastal rock or sediment, wave exposure, tidal range, and climate.

Coast, sea cliff, or dune According to strict definition, the coast is the strip of land affected by the ocean, extending landward from the shore or beach to the first major change in terrain (Figure 5–3). In common usage, the coast includes the shore as well as the landward zone. The coastline lies at the landward edge of the beach and forms the boundary between the shore and the coast. The shoreline at the seaward edge of the beach is the boundary between the ocean and the shore. On a rocky coast that has a beach at the base of a cliff, the coastline is marked by the cliff. On a barrier island, the coastline is formed by the seaward edge of the dune field and the coast includes the dune and the remainder of the barrier island. Along a low-lying, marshy coast, the coastline is indicated by the first appearance of permanent land vegetation.

Shore or beach The shore is a narrow strip of land in immediate contact with the sea, which lies between high and low water (Figure 5–3). A shore of loose or unconsolidated material, such as sand or gravel, is called a beach. A beach is usually divided into a foreshore, which slopes toward the sea, and a backshore, which is nearly horizontal or slopes toward the land. The backshore extends from the base of the cliff to the break in slope, marking the foreshore. The berm is a flat area on the backshore, covered by sand or gravel. Some beaches have no distinct berms and others have several. A beach scarp is a small, nearly vertical cliff cut into the beach by wave erosion. Beach scarps are usually less than a meter high, but higher scarps can be found.

The foreshore, or beach face, is the seaward sloping portion of the beach between the backshore and the ordinary low-water level (Figure 5–3). On a relatively flat beach covered by fine sand, the foreshore may include the entire beach, from the low-water line to the base of the cliff. For example, at South Beach, Oregon, on the central Oregon coast, the broad, flat foreshore is more than 400 meters wide (Figure 5–6). A steep foreshore forms when the beach is covered with coarse sand or gravel. The berm crest marks a sharp break in slope between the foreshore and the backshore on a steep beach. During severe storms, sand carried up the foreshore and over the berm crest is deposited on the backshore. A steep foreshore is often marked by beach cusps that are crescent-shape depressions, separated by pointed projections of sand.

Inshore or surf zone The inshore or surf zone extends seaward from the foreshore to just beyond the breaker zone. The inshore zone commonly includes a series of longshore bars and troughs or rip channels (Figure 5–3). The longshore bars are roughly parallel to the shoreline and are separated by deeper troughs. Large waves often break over the bars, transporting water and sediment toward the shore. Longshore currents flow parallel to the shore in the troughs between the bars, and between the foreshore and the inner bar. Rip currents are formed where the longshore currents are turned seaward and cut across the bars forming rip channels.

Ridge and runnel refers to longshore bars and troughs that form in the inshore zone and move onto the foreshore. When the ridge develops, it is usually completely submerged at low tide in the inshore zone. However, as sand is carried over the leading edge of the ridge by wave action, the bar slowly moves shoreward and is exposed at low tide. Within a few weeks the ridge moves across the foreshore and becomes welded to the berm.

Some longshore bars that form in deeper water are more or less permanent and do not migrate toward the shore. Many bars maintain the same approximate position for several months or even years.

Offshore or continental shelf

The offshore zone, or continental shelf, extends from the seaward edge of the breaker zone to the edge of the shelf. The offshore zone is usually relatively flat and is only affected by waves during severe storms. Where loose sediment is present in the offshore zone, ripple marks are often an indication of weak currents or long period swells. Linear mounds of sand in the offshore zone often represent relict bars that were formed by wave action during lower sea level stands.

The continental shelf off the Atlantic coast of the United States is a gently sloping plane that extends eastward about 100 to 200 kilometers to a depth of about 200 meters. On the Pacific coast of North America, the continental shelf is much narrower. In many places the shelf is broken into numerous blocks that form offshore islands, such as Catalina Island off the California coast.

Measuring beach profiles

Since many beach studies depend on recognizing changes in beach profiles, it is worthwhile spending a little time explaining how to survey a beach profile. In engineering studies where structures are to be constructed in the surf zone, a transit is used to make a detailed survey of the beach (Figure 5–4). However, in most beach erosion studies where repeated profiles are made across the same beach, the "stake and horizon" method provides a quick and easy surveying technique.

For a stake and horizon survey, the equipment consists of two wooden rods and a length of rope. The rodpersons measure the difference in elevation between the two rods, which are held at a constant separation by the rope, and the recorder keeps notes as they proceed down the beach. A reference stake is pounded into the sand as a starting point for the profile. One rodperson places his or her rod beside the reference stake and measures the height of stake above the sand. The second rodperson stretches the rope horizontally from the first rod along a

FIGURE 5–4. A transit is used to survey a beach profile at Newport, Oregon. Driftwood logs are brought in by the winter storms.
Photo by W. T. Fox.

line perpendicular to the shore. Then, the first rodperson sights from his or her rod across the top of the second rod to the horizon (Figure 5–5). By reading along the line of the horizon on his or her rod, the first rodperson measures the difference in elevation between the two rods.

When the first reading is complete, the first rodperson walks down the beach past the second rodperson, pulls the rope taut, and lines up his or her rod with the first rod and the reference stake. Then, the second rodperson takes a reading over the top of the first rod to the horizon. To complete the profile, the rodpersons leapfrog down the beach until they are chest-deep in the water. It usually takes about 15 minutes to half an hour to complete a profile, depending on the width of the beach and the speed of the crew. When the profile is complete, the data are plotted on a graph to get a picture of the profile.

Surf Zone Processes

Coastal processes including waves, tides, and nearshore currents are responsible for determining the width and shape of the coastal zones. Shoaling waves are the dominant force in eroding the coast and distributing sand and gravel along the shore. On

FIGURE 5–5. A rope and two calibrated rods are used to measure a beach profile. The difference in elevation is measured by sighting from the rear rod across the top of the forward rod to the horizon.
Photo by W. T. Fox.

the open coast, the major effect of tides is to raise and lower sea level, which spreads the effects of waves across a broader area. Near the mouths of inlets and estuaries, tidal currents transport the sediments, forming bars and tidal deltas.

Standing waves along the coast are responsible for many of the features in the surf zone and along the foreshore. Standing waves parallel to the shore determine the initial location and size of longshore bars. Standing waves, or edge waves at right angles to the coast, control the spacing of beach cusps on the foreshore and rip channels in the surf zone.

Nearshore currents are generated by waves that break along the shore. When waves approach the beach at an angle to the beach, longshore currents move along the beach parallel to the shore. At even intervals along the beach, longshore currents cut across the bars and head seaward, forming rip currents.

Shoaling and breaking waves

Storm waves and swells moving across the continental shelf start to shoal as they feel bottom (see Chapter Three). Initially, the wave length and height are decreased until the water depth is less than one-sixth the wave length (Figure 3–9). As the waves approach the surf zone, the length continues to decrease, but the height starts to increase. In the surf zone, the steepness ratio (height/length) increases until the waves become unstable and break. After a wave breaks, it moves toward the shore as a bore that carries water and sand up the foreshore.

Breaking waves in the surf zone. Wave behavior in the inshore zone depends on the bottom slope, and presence or absence of longshore bars. Where the bottom slope is relatively gentle (less than 2 degrees), large waves start to form, spilling breakers at the outer edge of the surf zone, and continue breaking until they reach the shore. Spilling breakers dissipate energy and are reduced in height as they move toward the beach. Where the bottom slope is steeper, but still relatively flat, the waves form plunging breakers when the wave height is 0.8 times the water depth. Most of the wave energy is dissipated by the plunging wave in the surf zone, and the remaining energy is released when the wave breaks on the beach.

Longshore bars. Where a gently sloping bottom is covered with loose sand in the surf zone, the sand is often moved by standing waves into a series of longshore bars (Figure 5–6).

FIGURE 5–6.
Beach, longshore bars, and rip channels exposed at low tide at South Beach, Oregon.
Photo by W. T. Fox

Spacing and depth of the bars are controlled by standing waves in the surf zone, which will be discussed in the next section. Multiple sand bars along the shore act as a series of filters, reducing the height of the large waves before they break along the beach. For example, on the eastern shore of Lake Michigan, waves break over longshore bars that lie parallel to the beach. The crest of the inner bar is located 55 meters from the shore at a depth of 1 meter, the second bar is 90 meters from shore at a depth of 2 meters, and a third bar is 210 meters from shore at a depth of 3 meters.

During storms, waves break over the outer bar with a breaker height of about 2.5 meters. Some of the wave energy is dissipated as the wave breaks over the outer bar, and the height is reduced when the wave moves off the bar into the trough. When the wave advances out of the trough and over the second bar, the height once again increases and the wave breaks with a height of 1.5 meters. When the wave breaks over the inner bar, the height is about 0.7 meter, and when it finally reaches the beach, the breaker is a little over 0.4 meter high. As the waves break over each successive bar, some of the wave energy is converted to heat energy and turbulence, which lifts sand off the bottom.

Standing waves and longshore bars. Long period swells that arrive at regular intervals along the beach set up standing waves roughly parallel to the foreshore. Nodal and antinodal lines parallel to the beach are responsible for the spacing and depth of the longshore bars (Figure 5–7). The beach face acts as a reflecting wall, setting up the standing wave. At the antinode, the water motion is nearly vertical as the water moves up and down. Water motion at the nodes is essentially horizontal, sweeping sand toward the antinodes. Therefore, the sand accumulates under the antinodes, forming longshore bars that are parallel to the beach.

FIGURE 5–7. Standing waves with nodal lines parallel to the beach. Longshore bars form under the antinodes and troughs under the nodes.
Drawing by C. W. L.

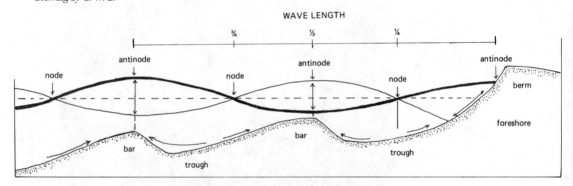

The spacing and depth of the longshore bars are directly proportional to the wave length of the standing wave. In deeper water at some distance from the beach, the standing wave becomes progressively longer (Figure 5–7). The bars formed at the antinodes are located at distances of one-fourth, three-fourths, and one and one-fourth wave lengths from the beach. Once the bars become established by standing waves, they become breaking sites for shoaling waves. Because bar crests are located at antinodes of the standing waves, the height of the standing wave is added to the shoaling wave to produce large breaking waves on the bars. The troughs between the bars are located at the nodes of the standing waves and, therefore, the shoaling waves over the troughs are not as high.

Edge waves and beach cusps. When waves are reflected off the beach face, standing waves known as edge waves are set up with the crests at right angles to the shore (Figure 5–8). In edge waves, the nodal and antinodal lines are aligned perpendicular to the beach face. Edge waves were first observed in

FIGURE 5–8. Edge waves are standing waves in which the nodal and antinodal lines are perpendicular to the foreshore. Beach cusps are formed at the antinodes where swash and backwash are greatest, and horns form at the nodes. *Drawing by C. W. L.*

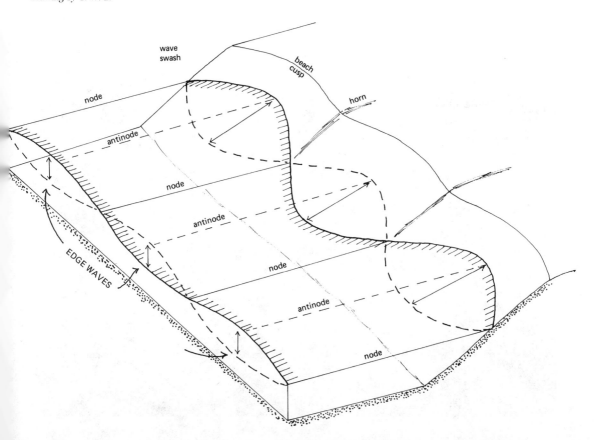

large wave tanks, and were thought to be caused by standing waves reflecting off the edges of the tank; hence, the name, edge waves. Edge waves have been recorded in the surf zone with sensitive wave recorders installed along the beach. Although the amplitude of the edge waves is usually quite low as compared with the incoming waves, they increase or decrease the size of the breakers. Edge waves are thought to be responsible for evenly spaced beach cusps along the berm crest, and for the regular spacing of rip currents along the beach (Figure 5–6).

Beach cusps develop along a moderately steep foreshore, with the horn of the cusp at the node of the edge wave and the embayment at the antinode (Figure 5–9). Therefore, one wave length of the standing wave includes a pair of cusps. The edge waves are formed by the incoming waves, and their crests are in phase. Along the foreshore, the higher swash alternates back and forth between each pair of cusps. When the crest of the standing wave is in phase with the crest of the incoming wave, the swash reaches higher up the beach (Figure 5–8). In the adjacent cusp, the trough of the standing wave is in phase with the crest of the incoming wave, and the swash does not reach as high. When the next wave reaches the beach, the crest and trough of the standing wave have switched positions in the cusps.

FIGURE 5–9. Swash marks and beach cusps on the foreshore at Plum Island, Massachusetts.
Photo by W. T. Fox.

Nearshore currents   Waves generate longshore currents that flow parallel to the beach in the trough between the inner bar and the beach, and the troughs between the longshore bars. When a wave breaks over a bar and moves toward the shore as a bore, water is carried into the trough between the bar and the shore. As the water piles up in the trough, it generates a current, parallel to the beach. The longshore current moves along the shore until it reaches a low point or break in the bar, where it turns seaward as a rip current. Near inlets, where tidal currents are flowing in and out between barrier islands, tidal currents are added to the longshore currents.

Longshore currents.   When waves approach the shore at an angle to the beach, a longshore current is formed and moves parallel to the shore (Figure 5–10). The strength of the longshore current depends on the height of the waves and the angle of wave approach. Wave refraction bends the crests of long period swells so that they are aligned almost parallel to the beach. Therefore, swells do not generate strong longshore currents. Storm waves that are steep and have shorter periods often approach the shore at a larger angle and produce the faster longshore currents. Longshore currents in excess of 2 meters per second have been measured during storms on Lake Michigan, but longshore currents from swells on the open ocean are generally less than 0.5 meters per second.

Longshore currents are dependent on the angle of wave approach and will reverse directions when the angle of wave approach changes. For example, on the east coast of Lake Michigan, longshore currents shift directions when warm and cold

FIGURE 5–10. When waves break at an angle with the beach, longshore currents are generated parallel to the shore. Offshore waves hit the beach similar to a broom hitting and sweeping the floor. *Drawing by C. W. L.*

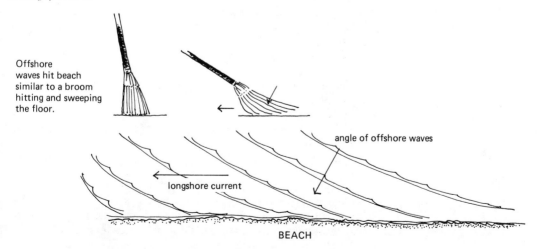

Offshore waves hit beach similar to a broom hitting and sweeping the floor.

angle of offshore waves

longshore current

BEACH

fronts pass over the coast. Following the passage of a warm front, the wind blows from the southwest and the longshore current flows to the north (see Chapter Two—Coastal Weather). When a cold front passes, the winds shift to the northwest and the currents flow to the south.

Several different methods are available for measuring longshore currents in the surf zone. Very accurate measurements for scientific studies can be obtained, using an electromagnetic current meter. As the water moves past the head of the current meter, an electric current is set up that is proportional to the speed of the longshore current. The electromagnetic current meter accurately records instantaneous changes in current velocity; however, a good electromagnetic current meter costs several thousand dollars.

If your budget is somewhat less but you are still interested in obtaining accurate results, a small digital flow meter can be used. When the digital flow meter is suspended in the current, the propeller spins and the number of revolutions is proportional to the speed of the current. The digital flow meter was originally designed for plankton flow nets, but it is rugged and well suited for work in the surf zone.

If your equipment budget is small, a current drogue may be used for obtaining current readings. Two fins are suspended beneath a float that moves along the surface with the longshore current. With two pairs of stakes at a known distance apart along the beach, the speed of the longshore current can be determined by timing the drogue as it moves with the current. Professor Jim Ingle at Stanford University has his oceanography students toss grapefruit into the surf to observe the nearshore currents.

Rip currents.   When waves break over nearshore bars, the excess water generates a longshore current, which eventually turns seaward forming a rip current. The rip current flows through a low point or saddle in the bar, forming a rip channel. When the rip current extends past the bar, it spreads out into a rip head.

Well-developed rip channels and nearshore bars are exposed along the Oregon coast at low tide (Figure 5–6). When water is carried over the bars by breaking waves, the current flows along the shore and heads out the rip channel. At low tide, the bars and rip channels are exposed and waves do not break across the bars. At mid-tide, bars are submerged and rip currents reach their maximum speeds. At high tide, waves break directly against the foreshore, and longshore currents flow parallel to the beach.

Large rip channels with longshore spacing of 15 to 1500 meters may be initiated by edge waves, but they are controlled by sand bars and rip currents in the nearshore zone. Large edge waves during storms concentrate high breakers along certain portions of the beach. When the alternating pattern of sand bars and rip currents becomes established by the edge waves, the features are maintained by the rip currents after the edge waves diminish.

Large, half-moon-shaped depressions, called lunate megaripples, are formed by the currents in the rip channels. Between mid and high tide, the rip channels are submerged and strong currents flow through the channels. At low tide, the water drains out of the rip channels and the lunate megaripples are exposed.

The water in the rip current moves seaward at 1 to 2 meters per second, which is faster than the average person can swim. Even strong swimmers have expired when they attempted to swim upstream against a rip current. If caught in a rip current, the best thing to do is to swim along the shore, for the rip is usually quite narrow. Along many California beaches where the rip currents are extremely dangerous, lifeguards mark the daily location of the rips with flags on the beach to warn the swimmers. Surfers who are knowledgeable often use rip currents as an easy passage seaward through the breakers to good surfing waves.

Beach Cycles and Erosion

Each wave that washes up on a beach either adds sand to the beach face or removes it. Minute by minute, hour by hour, and day by day, the beach profile is constantly modified by the waves. Although it is difficult to see changes that are taking place on a calm day, rapid changes during a storm are much more dramatic. Although cycles are present on all beaches, some beaches appear to be continuously eroding while others are accreting or building out.

The cycles in beach erosion and accretion are closely related to storm waves and swells. Steep waves generated by high winds during storms remove sand from the foreshore and cause beach erosion. The low, long period swells that break along the shore between storms restore the beach by moving longshore bars toward the shore.

In some areas, storms occur in all seasons of the year and beach erosion or accretion is related to individual storm cycles. Along other coasts, storms are concentrated at certain times of the year and produce a seasonal cycle of erosion and accretion. For example, on the coasts of Washington and Oregon, severe storms that come ashore during fall and winter strip sand from the beaches. During spring and summer, sand is returned to the shore within a series of large sand bars.

Coastal storms on the Great Lakes

Coastal storms have caused severe erosion problems on the Great Lakes. The eastern shore of Lake Michigan, at Holland and Stevensville, Michigan, has been studied in detail to develop a generalized model of erosion and deposition during a storm.

A series of beach profiles and maps of the nearshore area were surveyed before, during, and after the passage of several storms.

The severity of beach erosion along the Great Lakes depends on the frequency and intensity of the storms and the water level in the lakes. When the lake level was high, from 1952 to 1954, and from 1965 to 1970, the beaches, bluffs, and sand dunes were eroded, causing significant damage. For example, the bluff at Stevensville, Michigan, was eroded more than 7 meters (21 feet) during a single storm in July, 1969 (Figures 5–11A and B). When the lake level dropped, nearshore bars were exposed and sand was added to the shore, forming wide

FIGURE 5–11. The bluff at Stevensville was eroded back 7 meters during a storm in July, 1969: (A) beach before storm, (B) same beach after storm. A dark layer of magnetite was deposited on the beach during the storm. *Photo by W. T. Fox*

beaches. The broad beaches protect the bluff, and excess sand is blown from the beaches, forming protective dunes. Therefore, the cycles of erosion and accretion on the Great Lakes are closely related to lake levels, which are influenced by rainfall and drought conditions.

Storm cycles on the New England coast

Storm cycles on the coast of New England are closely related to northeasterly storms that move up the Atlantic coast. A typical "nor' easter" develops as a low-pressure center over the southeastern United States and swings northward along the coast. The strongest winds circulating counterclockwise around the storms blow from the northeast toward the New England coast.

From September, 1965, to April, 1969, the Coastal Studies Group at the University of Massachusetts measured more than 600 beach profiles at eight locations along the New England coast. From their studies, three basic profile shapes emerged, which are related to the occurrence of major storms and the recovery of beaches following the storms.

The early poststorm profile, that develops 3 or 4 days after the storm, is flat to concave upward with a generally smooth beach surface (Figure 5–12B). A sharply defined berm crest separates the foreshore from the backshore. A layer of garnet is often concentrated on the foreshore following the storm.

The early accretion phase lasts from 2 days to 6 weeks following the storm. A ridge and runnel system develops during the early accretion phase and advances slowly across the beach (Figure 5–12C). Often, a second ridge forms seaward of the primary ridge. Beach cusps are often present on the steep foreshore during the accretionary phase.

In the late accretion or mature phase, the ridge eventually grows in size and becomes welded to the berm (Figure 5–12E). Therefore, a new berm is formed and the beach builds out toward the sea. As waves wash over the top of the berm crest at high tide, layers of sand are deposited on the sloping backshore.

There are distinct summer and winter beaches on the New England coast. The beach profiles are related to the occurrence of storms, which can strike the coast during any season. However, hurricanes are more common during the late summer or early fall, and northeasters are most severe during the late fall, winter, and early spring.

Seasonal beach cycles on the Pacific coast

Beaches along the Pacific northwest are characterized by seasonal beach cycles. During the summer, the North Pacific High dominates weather patterns. Winds circulating in a clockwise direction around the high result in steady north winds along the coast. The north winds produce coastal upwelling and fog, but do not generate high waves along the coast.

In winter, frequent storms in the North Pacific are accompanied by high winds, large waves, and heavy rains. The wind direction shifts as storms move across the coast, but the dominant wind direction in winter is out of the southwest.

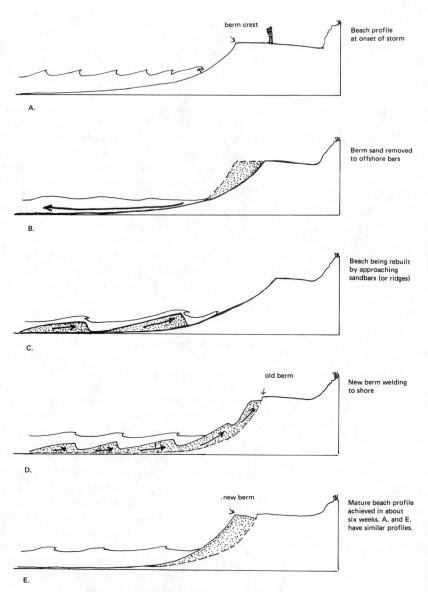

FIGURE 5–12. Beach profile (summer or winter) rebuilding after a storm.
Drawing by C. W. L.

A year-long study was conducted on the Oregon Coast from June, 1973, through May, 1974, to relate seasonal weather patterns to beach erosion cycles. The graphs of barometric pressure, longshore wind, and significant wave height clearly show the concentration of storms in winter (Figure 5–13). The wide range of barometric pressure from October through March indicates the passage of low-pressure centers or cyclonic storms. The plot of longshore wind shows the north wind, which pro-

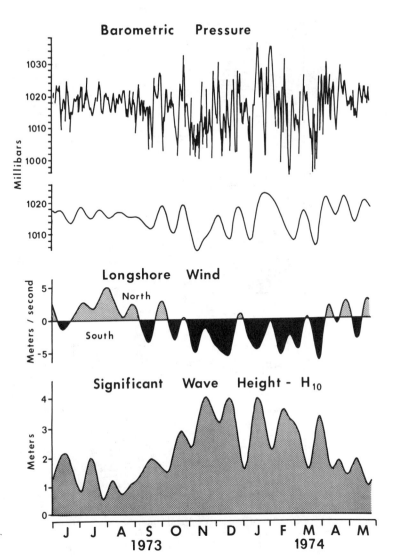

FIGURE 5–13.
Plots of barometric pressure,
longshore wind, and significant
wave height on the Oregon coast
show the increase of storm
activity during the winter months.
*From Fox and Davis, 1974, Office of
Naval Research.*

duced upwelling during summer, and the strong southwest winds during storms, which generated high waves during winter. The smoothed curve for significant wave height shows prolonged periods of 4-meter waves during the winter, with 1- to 2-meter waves during the summer. The waves reach heights of more than 7 meters several times during intense storms.

A section of beach 488 meters along the coast was surveyed twice a month for a year. The beach is located about a kilometer south of the jetty at Newport, Oregon. During July and August, 1973, several large sand bars moved onshore (Figure 5–14). The total sand volume added to the beach was 22,400 cubic meters. With the initial winter storms of October and November, 54,000 cubic meters of sand were eroded from the beach. Bare rock was exposed through the sand in several areas.

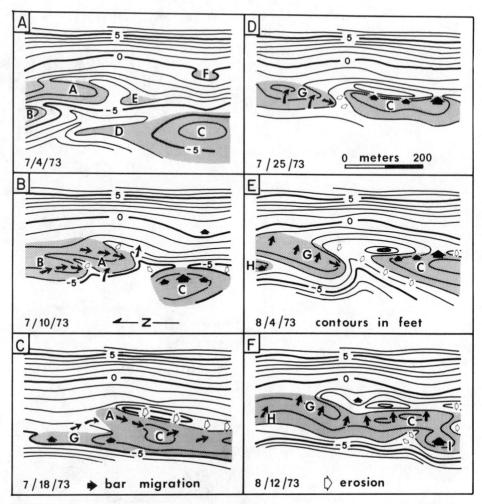

FIGURE 5–14. Sequence of maps showing bar migration and beach erosion at Newport, Oregon, July and August, 1973.
From Fox and Davis, 1974, Office of Naval Research.

The beach started to recover in December and January, when large bars formed on the foreshore. By May, 1974, the volume of sand on the beach was restored to the original volume of June, 1973.

During the winter, some of the sand was stored off the coast in the form of large longshore bars. Additional sand was added to the beach by a large landslide in January, 1974. The sand in the longshore bar was returned to the beach during the spring and summer of 1974.

The winter and summer beaches are typical of the California, Oregon, and Washington coasts. In some areas along the coast, the sand is stripped from the beach and rock is exposed along the shore. The sand returns to the beach in the summer, but the coastline has retreated due to landslides and erosion.

The average rate of coastline erosion along the central Oregon coast is about 2 meters per year.

Coastal erosion   Along many coasts, continuous coastal erosion is superimposed on storm cycles or seasonal patterns of erosion and accretion. In typical storm and seasonal cycles, sand that was removed from the beach during storms is returned to the beach during calm intervals between storms.

In coastal areas that face the brunt of frequent intense storms, the cliffs and dunes marking the coastline are eroded and the shoreline retreats. For example, between 1888 and 1958, the coastline on Cape Cod between Nauset Spit and Highland Light has been retreating at an average rate of about 1 meter per year. The soft cliffs at Suffolk, England, are eroding at an average rate of 3 to 5 meters per year. At Lowestoft on the Suffolk Coast, a 12-meter-high cliff of unconsolidated material was eroded back to 12.2 meters overnight. Where the cliff was only about 2 meters high, it was eroded back 27.4 meters by the same storm.

Beach erosion is very difficult to predict and almost impossible to stop. Beach erosion is quite variable because it depends on the strength of the cliffs or dunes, the intensity and frequency of the storms, and the exposure of the coast. Most attempts to prevent erosion are doomed to eventual failure, because the waves constantly batter and erode the man-made defenses.

Summary   A beach is a strip of sand or gravel along the shore that separates the land from the sea. The sediment on a beach was derived from the erosion of sea cliffs or was transported to the coast by rivers or glaciers. A gravel beach is covered by boulders, cobbles, pebbles, or granules, which come from weathered and broken rock fragments. Sand beaches are composed of fine grains of rounded quartz and feldspar.

A profile across the coast from land to sea includes four zones: coast, shore, inshore, and offshore. The coast is marked by sea cliffs, sand dunes, or permanent vegetation. The shore or beach between sea level and the coast is divided into foreshore, dipping toward the sea, and backshore, dipping toward the land. The inshore zone includes nearshore bars and rip channels in the surf zone. The offshore zone is below wave base and extends from the breaker zone to the edge of the continental shelf.

The shape and size of the beach are determined by breaking and standing waves in the surf zone. Nearshore bars are formed by standing waves parallel to the beach, which are reflected off the foreshore. Beach cusps are formed by edge waves, which are standing waves along the coast. Longshore currents, generated by waves that break at an angle to the shore, turn seaward to form rip currents.

Cycles in beach erosion and deposition are related to

coastal storms. The beach goes through several stages, from erosion during a storm to recovery by onshore migration of sand bars. Concentration of coastal storms in winter produces a summer and winter beach profile along many coasts.

Selected Readings

BASCOM, W. *Waves and Beaches*. Garden City, N.Y.: Anchor Books, Doubleday, 1980. A good introduction to beaches of the Pacific coast for the general reader.

DAVIS, R. A., JR. *Coastal Sedimentary Environments*. New York: Springer Verlag, 1978. A geological discussion of coastal environments, with a chapter on beaches.

DAVIS, R. A., JR. AND R. L. ETHINGTON. *Beach and Nearshore Sedimentation*. Tulsa, Okla.: Society of Economic Paleontologists and Mineralogists, Special Publication No. 24, 1976. An excellent collection of scientific papers dealing with several aspects of beach processes.

FISCHER, I. S. AND R. DOLAN. *Beach Processes and Coastal Hydrodynamics, Benchmark Papers in Geology*. Stroudsburg, Penn.: Dowden, Hutchinson and Ross, Inc., 1977. A collection of original geological papers on beach processes.

GIESE, G. S. AND R. B. GIESE. *The Eroding Shores of Outer Cape Cod*. Orleans, Mass.: Association for the Preservation of Cape Cod, Information Bulletin No. 5, 1974. A short explanation of coastal erosion problems on Cape Cod.

HAY, J. *The Great Beach*. New York: W. W. Norton, and Co., 1980. A naturalist's view of Cape Cod.

JOHNSON, D. W. *Shore Processes and Shoreline Development*. New York: John Wiley and Sons, Inc., 1919. An early book discussing the formation of beaches.

KAUFMAN, W. AND O. PILKEY. *The Beaches are Moving: Drowning of America's Shorelines*. Garden City, N.Y.: Anchor Books, Doubleday, 1979. A geologist's plea to stop development on beaches.

KING, C. A. M. *Beaches and Coasts* (2nd ed.). London: Edward Arnold Ltd., 1972. An excellent examination of beach and coastal processes.

KOMAR, P. D. *Beach Processes and Sedimentation*. Englewood Cliffs, N.J.: Prentice-Hall, Inc., 1976. A well-written advanced text on beach processes for geologists and engineers.

KOPPER, P. *The Wild Edge, Life and Lore of the Great Atlantic Beaches*. New York: Times Books, 1979. A lively and exciting book by a naturalist writer, with a chapter on beaches.

OGBURN, C. *The Winter Beach*. New York: William Morrow and Company, 1966. An interesting narrative of a naturalist's trip down the Atlantic coast during winter.

SHEPARD, F. P. *Submarine Geology* (2nd ed.). New York: Harper and Row, 1972. In-depth treatment of beaches, with emphasis on the California coast.

ZENKOVICH, V. P. *Processes of Coastal Development*, T. A. Steers (ed.). New York: John Wiley and Sons, 1967. A Russian's view of beach and coastal processes.

Cedar Island – Georgia   aerial view   CWL

SIX

Barrier islands and spits

Introduction to Barrier Islands and Spits

Many of the most beautiful sand beaches of the world are found on barrier islands and spits. Along the east coast of United States, a string of barrier island extends from Plum Island, Massachusetts, to Padre Island, Texas. The barrier islands are separated from the mainland by a salt marsh or lagoon. On the Pacific coast, barrier islands are present on the south coast of Alaska and on the west coast of Mexico. Although there are no barrier islands in Washington, Oregon, or California, several spectacular sand spits extend across the mouths of rivers and estuaries. The sand supplied by the rivers and eroded from the coast is carried along the coast by waves and currents to form the spits.

Several of the barrier island and spits have been preserved as National Seashores. Each National Seashore represents a different type of barrier island or spit, so they provide excellent opportunities for study and enjoyment. Nauset Spit and Monomoy Island are examples of spits and islands formed from glacial sediments on Cape Cod, Massachusetts. Fire Island National Seashore is a barrier island along the Great South Bay on the south shore of Long Island, New York. Assateague Island in Maryland and Virginia is a good example of an undisturbed barrier island along the mid-Atlantic coast. Cape Hatteras National Seashore on the Outer Banks of North Carolina is separated from the mainland by Pamlico Sound. Cumberland Island National Seashore is one of the sea islands on the Georgia coast. Canaveral National Seashore, north of the John F. Kennedy Space Center, is typical of the converging spits that form capes along the Florida coast.

In the Gulf of Mexico, several barrier islands occur along the Mississippi coast in the Gulf Islands National Seashore. Padre Island National Seashore, including Padre and Mustang Islands, stretches from Corpus Christi to Brownsville, Texas.

On the Pacific coast, an arcuate spit curves across the bay at Point Reyes National Seashore, a short distance north of San Francisco. Several large spits are present along the coasts of Washington and Oregon. The best example is Long Beach, which reaches north from the mouth of the Columbia River across Willapa Bay.

Description of Barrier Islands

Long, dune-covered islands that are roughly parallel to the coast and are separated from the mainland by a shallow lagoon or salt marsh are known as barrier islands. Barrier islands often have shallow bars on a gentle nearshore slope that produce a broad surf zone. At low tide, most of the wave energy is expended as waves break across the surf zone. At Padre Island, Texas, where the barrier island is covered by fine sand, the foreshore is relatively flat and broad. However, when coarse sand covers the foreshore, such as is found on Siletz Spit, Oregon, the beach face is steep with a prominent berm crest and backshore.

A dune or series of dunes is present on many barrier islands where wind-blown sand is trapped by beach grass. The beach grass thrives when it is buried, and grows upward through the sand, anchoring the dune. Where the dunes are high, a dense thicket or maritime forest becomes established in the low swale between dunes or in the protected area behind the dunes. The Sunken Forest on Fire Island is an example of a maritime forest. When the dunes migrate toward the mainland, the forest or thicket is buried by the advancing sand, and may reappear on the seaward edge of the dunes.

On the lagoon side of the barrier or spit, a salt marsh develops on low, intertidal flats. Grasses that grow in the salt marsh are adapted to daily submergence by the tides. Salt water pours into the lagoon with the flooding tide, and is flushed with the ebb. Small tidal creeks feed into large channels that drain the salt marshes. Freshwater rivers and streams from the mainland flow into the lagoon, producing brackish water where the fresh and salt water mix. The lagoon usually has a normal or above-normal marine salinity near the tidal inlet, but salinity decreases in the upper reaches of the tidal creeks.

The soft muds and sands of the salt marsh and tide flats support a diverse benthic fauna and flora. The protected waters of the lagoon provide an ideal nursery for many varieties of shellfish and finfish, which spend most of their adult lives in the open sea. The distribution of organisms and environments related to the barrier islands are discussed in Chapter 7.

The Origin of Barrier Islands

Several different explanations have been given for the origin and development of barrier islands. Three major hypotheses are used to account for different barrier islands: (1) the deBeaumont-Johnson concept of the emergence of submarine bars; (2) the Gilbert-Fisher idea of spit growth and later breaching by inlets; and (3) the McGee-Hoyt hypothesis of beach ridge submergence. Each hypothesis was originally proposed in the 1800s and later reexamined. For the wide variety of barrier island types, the three processes, either alone or in combination, can explain the origin of almost all barrier islands.

Early theories

Emergence of submarine bars—Elie deBeaumont. In 1845, a French geologist, Elie deBeaumont, studied barrier islands from maps of Europe, North America, and North Africa. He concentrated on examples from the North Sea and the Gulf of Mexico to support his concept of bar emergence for the origin of barrier islands. According to deBeaumont, waves sweeping across the shallow bottom picked up loose sand and deposited

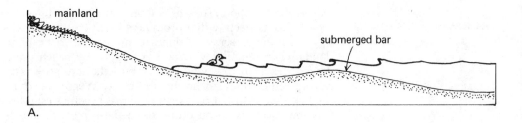

A.

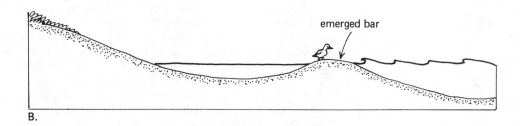

B.

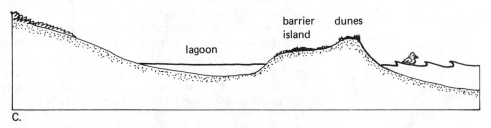

C.

FIGURE 6–1. Emergence theory of barrier island formation proposed by deBeaumont in 1845. (A) Waves breaking over submerged bar; (B) bar emerges above sea level; (C) bar develops into barrier island and lagoon.
Drawing by C. W. L.

it on submarine bars. As sand was added, the bars eventually rose above sea level to become barrier islands (Figure 6–1).

Spit growth—Karl Gilbert. In 1885, Grove Karl Gilbert proposed that barrier islands were formed from spits by currents flowing parallel to the coast. Individual islands were created when the spits were breached by waves during storms (Figure 6–2). Gilbert's hypothesis is based on Lake Bonneville, a large Pleistocene lake that covered more than 55,000 square kilometers in western Utah. The lake reached its fullest extent when rivers from heavy rains and melting glaciers flowed into the lake. Great Salt Lake is all that remains of the larger Lake Bonneville, but earlier bars and spits are preserved on terraces etched in the Wasatch Mountains, east of Salt Lake City.

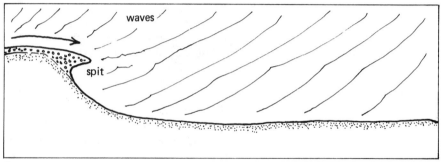

A.

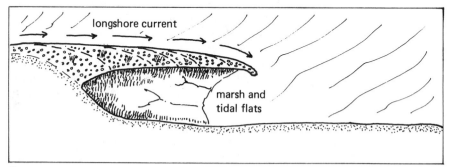

B.

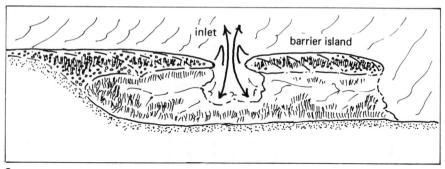

C.

FIGURE 6–2. Spit theory of barrier island formation proposed
by Gilbert in 1885. (A) Spit starts to grow from point of land;
(B) spit is extended along the coast by longshore currents;
(C) spit is breached during a storm, forming a
tidal inlet and barrier island.
Drawing by C. W. L.

Drowning of the coastal area—W. D. McGee. W. D. Mc-
Gee suggested in 1890 that barrier islands were produced by the
drowning of the coastal area during a rise in sea level. He used
vivid language to dramatize the advance of the sea: "They are
wave-fashioned plains, but recently wrested from the ocean,

155

and ocean reclaims its own. Already its octopus arms have seized the lowlands in its horrid embrace, and day by day, month by month, year by year, generation by generation, the grasp is tightening, the monster creeping further and further inland, each year the water march advances a rod." McGee considered the drowned river valleys along the eastern United States as evidence of submergence, and assumed that barrier islands were partially submerged beach ridges and dunes that lined the coast. He refers to the barrier islands as keys, based on the Florida Keys that extend southeast from the tip of Florida.

Bar emergence reexamined—Douglas Johnson. In 1919, Douglas Johnson reviewed the evidence and supported the concept of bar emergence originally put forth by deBeaumont. Johnson finished his classic book, *Shore Processes and Shore Line Development,* while he was aboard a troop ship bound for France. Johnson discarded McGee's hypothesis as inadequate to account for the wide distribution of barrier islands. In an attempt to decide between the emergence idea of deBeaumont and the spit hypothesis favored by Gilbert, Johnson decided to

FIGURE 6–3. Douglas Johnson tested the theories of bar emergence and spit formation by plotting nearshore slope. (A) An island formed by spit, profile intersection at coast; (B) an island formed from emerged bar, profile intersection is on the mainland.

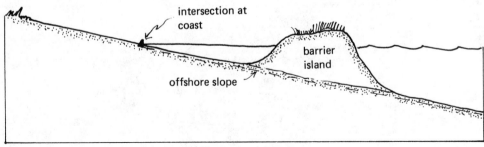

A. Island formed from spit sits on top of bottom profile

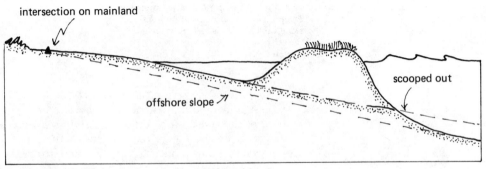

B. Bottom profile scooped out seaward to form island

test the hypotheses by drawing profiles across a series of barrier islands. The task of constructing the profiles was assigned to Miss B. M. Merrill, one of Johnson's graduate students at Columbia University. Johnson reasoned that the offshore slope would intersect the mainland at the coast if the islands were formed by the extension of spits due to longshore currents. However, if the barriers were formed by the buildup of submarine bars, the ocean floor seaward of the islands would be scooped out by the waves, to provide sand for the islands. Therefore, if the slope of the ocean bottom was projected shoreward, it would intersect the land surface well inland of the mainland coast (Figure 6–3). For emerged bars, the ocean floor just seaward of the island would be concave upward. The profiles drawn by Miss Merrill favored the formation of barrier islands from the emergence of offshore bars, and laid the argument to rest for several decades.

Later revisions of barrier island theories

In the 1960s, barrier islands were once again intensively studied, this time by a group of geologists sponsored by the American Petroleum Institute. The geologists were interested in the formation of barrier islands so that they could interpret the origin of buried sand deposits filled with petroleum. With muds containing marine fossils on one side and lagoonal muds on the other, the ancient barrier islands provided excellent reservoirs for oil. Francis Shepard from Scripps Institution of Oceanography cast doubt on the evidence gleaned from Johnson's profiles and claimed that it would be difficult to favor either the spit or emergence hypothesis, and that both were probably important. Shepard emphasized that barrier islands grew upward between bay muds on one side and shelf muds on the other.

Submergence of beach ridge—John Hoyt. In 1967, John Hoyt wrote a paper discussing barrier island formation, which reopened the controversy once again. Hoyt was struck by the absence of marine organisms or beach sediments on the mainland shore of the lagoon, or preserved as fossil material beneath the sediments of the lagoon. He reasoned that if a barrier island was formed by emergence of an offshore bar, marine fossils should be present on the old beach that existed before the bar emerged. Similarly, if a barrier island formed from a spit that expanded along the shore and was cut off by the tidal inlets, the original marine beach should be preserved along the mainland coast. Because marine fossils and beach deposits are lacking along the mainland side of the lagoon, Hoyt concluded that the barrier island must have been formed by the submergence of a coastal dune field or beach ridge. When sea level rose during the retreat of the glaciers, the area behind the ridge was flooded, forming a lagoon, and the ridge was left behind as a barrier island (Figure 6–4). Once the barrier island has been formed, waves and longshore currents mold the islands to their present shape.

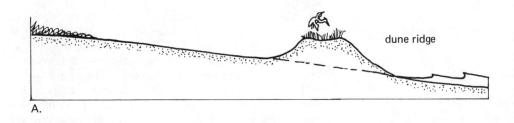

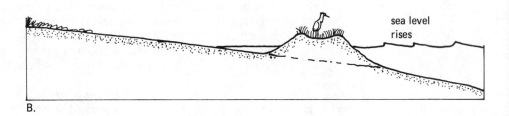

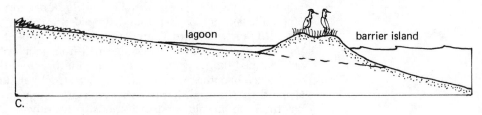

FIGURE 6–4. Barrier formation by beach ridge formation proposed by John Hoyt, 1967. (A) Dune ridge along coast; (B) rise in sea level starts to submerge dune; (C) barrier island and lagoon formed from ridge.

Spit formation—John Fisher. In 1968, Fisher supported Gilbert's idea of spit formation of barrier islands. Fisher agreed with Hoyt that the lack of marine fossils along the mainland beach argues against the formation of barrier islands by bar emergence. However, he pointed out that soils or forests that grew behind the beach ridge should be preserved beneath the lagoonal muds if the island was formed by a submerged ridge. Since land-derived sediments are not found beneath the lagoonal muds, Fisher concluded that the islands were created from breached spits. He also pointed out that the long, straight barrier islands would be difficult to form from the submergence of an irregularly shaped coastline.

Multiple origin of barrier islands—Maurice Schwartz. In 1971, Maurice Schwartz attempted to draw together the diverse origins of barrier islands into a unified concept that accepted the multiple origin of barrier islands. He stated that there are two types of barrier islands: primary barrier islands, which are formed on land that is later flooded by the sea, and secondary barrier islands, which developed seaward of the primary coast on the continental shelf. The primary barrier islands include the submerged beach ridges that originally formed as dunes or beach ridges on the shore. The secondary barrier islands would include two types: the breached spits and the emergent offshore bars. Schwartz includes a third category of composite spits for those that were formed by a combination of two or more of the methods just described.

Erosion and Migration of Barrier Islands

Once a barrier island has formed, it may remain stationary, retreat toward the mainland, migrate along the coast, or prograde toward the ocean. The seaward edge of the island is often eroded by waves and longshore currents. During large storms and hurricanes, storm surges and waves pass through breaks in the dunes and deposit sand in washover fans behind the dunes.

Sand eroded from the beach is carried toward the inlets by longshore currents. The sand moves in and out of tidal inlets, forming ebb and flood tidal deltas. On flood tide, wave-generated longshore currents combine with tidal currents to move sand into the inlet. Some of the sand is returned to the beach by the ebb tide, but a large portion remains to fill in the lagoon. With the combination of beach erosion and filling of lagoons, islands retreat toward the shore.

When a large supply of sand is available to the barrier islands from rivers, the islands build or prograde seaward over the nearshore sediments. Galveston Island on the Texas coast near Houston has prograded several kilometers since its formation.

Tidal range

Tidal range is an important factor in determining whether washover fans or tidal deltas will dominate barrier island migration. On a microtidal shoreline (tidal range less than 2 meters), washover fans are the dominant method of barrier island migration. For example, along the Texas coast where the tidal range is less than a meter, a single low dune ridge forms behind the beach. During hurricanes, the dune is easily breached by large waves and sand is carried through breaks in the dune line.

On a mesotidal coast (tidal range is between 2 and 4 meters), tidal inlets provide a passageway for sand into and out of lagoons. With the increased tidal range, the beach is wider and more sand is available for the dunes. In the northeastern states, dunes are generally higher and often there are multiple lines of dunes parallel to the beach. Where longshore current is predominately in one direction, spits wrap around the end of the island, forming recurved sets of beach ridges. Sand carried into an inlet with the rising tide is deposited on the flood tidal delta. With the falling tide, sand moving seaward with the ebb currents merges with longshore bars to form the ebb tidal delta.

FIGURE 6–5. Tidal inlet between barrier islands at Matanzas
Inlet, Florida. Flood delta is landward of bridge and ebb delta
is seaward of bridge. Flood channels enter inlet on each side,
and ebb channel flows out on the left margin. Breaking waves
mark the edge of the terminal lobe on the ebb delta. Ebb
shield is on inner edge of flood delta.
Photo by W. T. Fox.

Tidal inlets Tidal inlets that form between barrier islands are important for
the process of barrier island migration. The inlet is accompanied by two major bodies of sand, the ebb tidal delta seaward
of the inlet, and the flood tidal delta on the landward side. Usually, the flood delta is much larger than the ebb delta because
sand on the ebb delta is dispersed by waves and longshore currents. Ebb and flood tidal deltas are well displayed at Matanzas
Inlet near Saint Augustine, Florida on the Atlantic coast (Figure
6–5).

    Ebb tidal deltas. The ebb tidal delta projects seaward
from the inlet throat at Matanzas Inlet (Figure 6–6). The ebb
current flows seaward in the ebb channel, which is constricted
in the inlet throat. The ebb channel flairs out like the horn of a
trumpet on the ebb delta. The interaction of the ebb tidal currents and wave-generated longshore currents produces a broad
sheet of sand seaward of the inlet, known as the swash platform. Swash bars are built by waves and currents on top of the
swash platform. The terminal lobe is a shallow bar that wraps
around the swash platform and marks the seaward limit of the
ebb tidal delta (Figure 6–6).

160

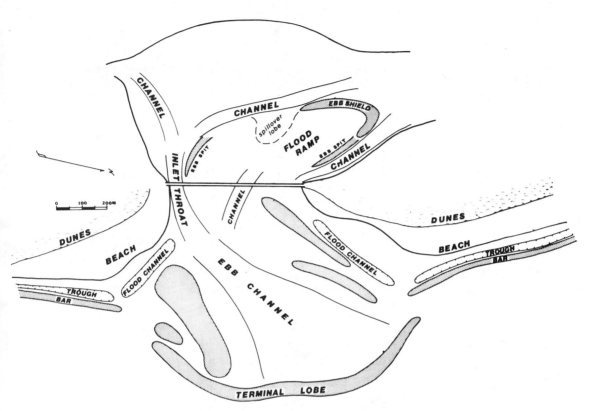

FIGURE 6–6. Major morphologic features of Matanzas
Inlet, Florida.
From Davis and Fox, 1980.

Longshore bars and troughs are present north and south
of the tidal inlet (Figure 6–6). During flood tide, the longshore
current merges with the flood current and enters the inlet
through the marginal flood channels. The flood current is directed onto the flood tidal delta on the landward side of the
highway bridge.

Flood tidal deltas. Flood currents flow up the flood ramp
that forms the flood tidal delta at Matanzas Inlet (Figure 6–6).
The flood ramp becomes progressively shallower approaching
the ebb shield. The ebb shield is a lip of sand on the margin
of the flood ramp that is exposed above the water at low tide. It
is called the ebb shield because it diverts the ebb current into
channels on either side of the flood ramp. A spillover lobe is
formed where ebb currents flow over the ebb shield and deposit
sand on the flood ramp. The ebb current also forms an ebb spit
where it flows along the edge of the flood ramp. The flood ramp
looks like the carapace or shell from a horseshoe crab that has
been flipped on its back. The outer rim of the crab corresponds
to the ebb shield, and the trailing edges of the crab shell would

be the ebb spits. The flood current would flow from the tail over the leading edge of the crab.

Sand waves and megaripples. Tidal currents leave their impression in the sand on the surface of the tidal delta. As tidal currents flow across a delta, sand is swept into giant ripple marks that are known as sand waves and lunate megaripples. The ripples, which are exposed at low tide, make beautiful patterns on the bottom.

Sand waves and megaripples can be used as an indication of the speed and direction of the current that formed them. For example, if the water is about 30 centimeters (1 foot) deep, small ripples will form when the current speed reaches about 22 centimeters per second. When the current speed is increased to about 35 centimeters per second, the ripples increase in size and become sand waves. Sand waves are generally 20 centimeters to a meter high and 3 to 20 meters long when measured from crest to crest. The sand waves have fairly straight crests with a gently sloping backside and a steep leading edge (Figure 6–7).

FIGURE 6–7. Small current ripples are superimposed on large sand waves at Matanzas Inlet, Florida. Sand waves have a wave length of 3 to 5 meters and a height of ½ meter. Ebb current flowed from left to right.
Photo by W. T. Fox.

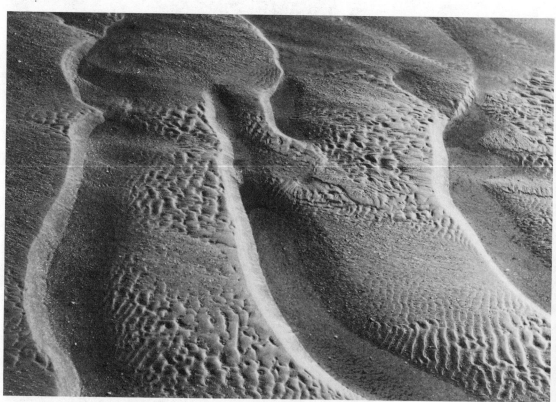

FIGURE 6–8. Lunate megaripples with small current ripples
at Matanzas Inlet, Florida. Megaripples are about 3 meters
long and 30 centimeters high. Ebb current flowed to
the right.
Photo by W. T. Fox.

When the current speed is increased to about 45 centimeters per second, the sand waves are transformed into lunate megaripples (Figure 6–8). The megaripples have a scalloped crest with a deep plunge pool downstream from the megaripple. The water appears to boil or bubble when it passes over the crest of the ripple into the plunge pool. The lunate megaripples are about 20 to 50 centimeters high and 3 to 5 meters long.

Washover fans When a wave surge passes through a dune line and spreads out on the salt marsh or tidal flat, sand is deposited in a washover fan (Figure 6–9). On the Atlantic coast, overwash usually occurs during a nor'easter or a hurricane, when waves and storm surge are combined with high spring tides. Overwash can occur through narrow gaps in high dunes, or across wide sections of the barrier island where the dunes are low or absent.

If the dune line is fairly continuous and relatively high, storm waves will not penetrate the dunes very often and overwash will be infrequent. If overwash is prevented by building up the dunes, beach erosion continues and the slope of the foreshore is increased. The sand that was eroded from the foreshore

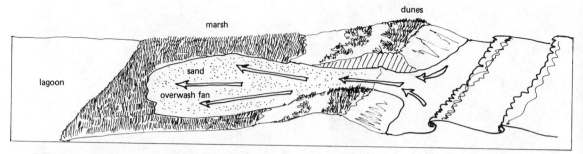

FIGURE 6–9. Washover fan is formed when a wave surge
passes through the dune line and spreads sand on the marsh.
Drawing by C. W. L.

is deposited on offshore bars or carried along the beach to form
spits at the end of the island.

Overwash can occur wherever the dune line has been low-
ered by natural or artificial causes. Where the protective covering of beach grass is disturbed, wind will remove the sand,
forming a blowout. Footpaths and dune buggy trails often follow natural low breaks in the dunes. When grass is beaten down
or removed, the low break is enlarged, providing a natural passage for a washover channel. Wave surges passing through the
channel tend to erode the dunes on each side and broaden the
channel.

Outer banks of North Carolina. On the Outer Banks of
North Carolina, where the tidal range is low and storm frequency is relatively high, overwash is a fairly common occurrence. Large washover fans display a light-colored sand on top
of the tidal marshes (Figure 6–9). The marsh grass rapidly
grows through the washover fan and incorporates the sediment
into the salt marsh.

When attempts are made to prevent overwash by planting
beach grass on the dunes or building seawalls along the base of
the dunes, erosion of the beach face accelerates. The slope of the
foreshore becomes steeper and the width of the backshore is decreased, reducing the size of the beach. For example, on Cape
Hatteras National Seashore, the dunes were planted with beach
grass to trap sand and increase the size of the dunes. The higher
dunes prevented overwash and the width of the beach was reduced by storms. When water was piled into the lagoon by
storm surges, the barrier island was also eroded from the lagoon side. When the National Park Service was made aware of
the problems associated with dune restoration on Cape Hatteras, the policy was reversed. The dunes are no longer artifi-
cially maintained, and overwash is once again permitted.

New England coast. In New England, where tidal range
is high and beaches are backed by large dunes, overwash is a

164

fairly infrequent event. During exceptionally large storms and hurricanes, overwash does occur, but washover fans are not common. In the northeast, the marsh grass does not respond rapidly to burial by washover fans, and it takes several years for a new crop of marsh grass to cover the washover fan. The response of different grasses to burial will be further discussed in Chapter Seven.

Hurricane washover in Texas. If the washover channel is cut deep enough through the dunes, it may erode below sea level and produce a tidal inlet. Along the Texas coast where barrier islands are low and hurricanes are frequent, several washover channels have become inlets. During hurricane David in August, 1979, forty new inlets were opened along the Texas coast.

During a hurricane, storm surge piles excess water into the bay. If a washover channel is cut below sea level by waves, the channel may be enlarged into a tidal inlet when excess water escapes from the bay. The volume of water that normally moves into and out of a lagoon through a tidal inlet is known as the tidal prism. Usually a barrier island-lagoon system maintains a relatively constant number of tidal inlets to handle the volume of the tidal prism. If the number of inlets is increased, the volume of water moving through each inlet will decrease, and sediment will fill in the inlet mouth. The smaller inlets are rapidly filled with sand and closed while larger inlets maintain the tidal prism.

Distribution of Barrier Islands and Spits

World distribution

Between 12 and 13 percent of the world's coastlines contain barrier islands or spits. About 76 percent of the barrier islands are located on trailing edges of spreading continents, such as North America and Africa, or along the coasts of marginal seas behind island arcs (see Chapter One). Broad continental shelves on the trailing edges of continents are conducive to the growth of barrier islands.

Table 6–1. DISTRIBUTION OF BARRIER ISLANDS ALONG THE BORDERS OF CONTINENTS.

North America	33.6%
Asia	22.7%
Africa	18.7%
South America	10.3%
Europe	8.4%
Australia	6.8%

More than one-third of the coast of North America is bordered by barrier islands (Table 6–1). The islands extend from the Gulf of Saint Lawrence through the Gulf of Mexico to the

Yucatan Peninsula. Large spits along the Pacific coast resemble barrier islands, and some actual barrier islands are found in Baja, California, and along the southeast coast of Alaska.

More than one-fifth of the coast of Asia is flanked by barrier islands. Some of the best-developed barrier islands are found on the coast of China along the Sea of Japan. In Africa, less than one-fifth of the coast has barrier islands, with the best examples found on the Durbin coast of South Africa along the Indian Ocean. Barrier islands also form on the Nile delta along the Mediterranean Sea.

Barrier islands are less common along the coasts of South America and Australia. In South America, about one-tenth of the coast has barrier islands, with some of the world's largest barrier islands found on the southeast coast of Brazil. In Australia, about 7 percent of the coast has barrier islands, with unusual names such as the Coorong and East Gippsland Barriers. Ninety Mile Beach, east of Melbourne, is one of the longest continuous barrier islands in the world.

In Europe, only 8 percent of the coast has barrier islands, with the greatest concentration of barriers flanking the low countries along the North Sea. Along much of the coast, dikes have been constructed across the mouths of the tidal inlets to prevent flooding during severe storms. Windmills were used to pump sea water from the lagoons to form polders, which include farmland and cities reclaimed from the sea by dikes. Much of the reclaimed land of the Netherlands lies below sea level and would be subject to extensive flooding if the dikes were breached.

Barrier islands and spits of North America

Barrier islands are present along the east coast of the United States from Maine to Texas. However, the different local geology, sediment supply, wave conditions, and tidal range have produced a variety of barrier islands along different parts of the coast. It is convenient to discuss the barrier-island types by regions, starting with New England, moving southward to Florida and into the Gulf of Mexico. Next, the barrier islands or spits that are present on the Pacific coast of North America will be considered.

New England coast. In New England and New York, the barrier islands were developed by the reworking of large glacial deposits. On Cape Cod, for example, beaches were formed as large spits that extended from the mainland (Figure 6–10). Several lobes of ice moved down from the north to form Cape Cod. Glacial moraine and outwash plains formed in front of the glacial lobes when the ice stopped advancing. Martha's Vineyard and Nantucket Island were formed from glacial moraine at the farthest advance of the ice. When the glaciers retreated slightly, the main body of Cape Cod was formed by a recessional moraine from the ice lobe in Cape Cod Bay. A long ridge of glacial

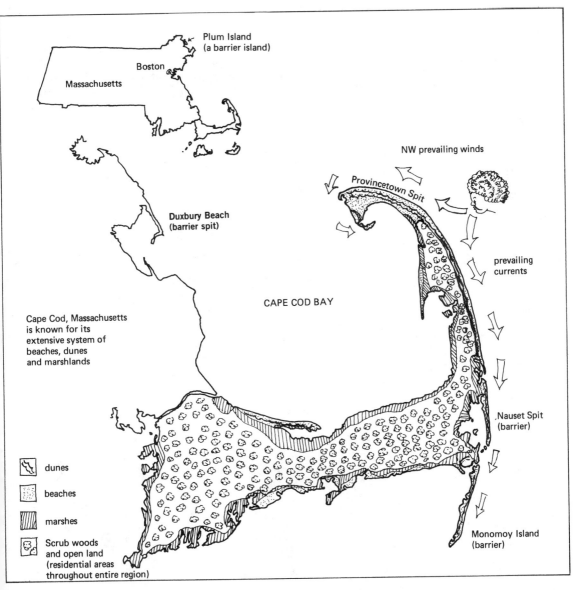

Plum Island
(a barrier island)

Boston

Massachusetts

NW prevailing winds

Provincetown Spit

Duxbury Beach
(barrier spit)

prevailing
currents

CAPE COD BAY

Cape Cod, Massachusetts
is known for its
extensive system of
beaches, dunes
and marshlands

Nauset Spit
(barrier)

dunes

beaches

marshes

Scrub woods
and open land
(residential areas
throughout entire region)

Monomoy Island
(barrier)

FIGURE 6–10. Geological features of Cape Cod, Massachusetts.
Drawing by C. W. L.

deposits extended to the north between two ice lobes to form the forearm on the Cape. When glaciers melted to the north, waves from the Atlantic eroded unconsolidated sands and gravels in the glacial moraines. The sand was moved along the shore by currents to form large spits. The Provincelands at the north end of Cape Cod were built out by the northward-flowing longshore currents. The currents split and Nauset Spit built out to the south, blocking off Pleasant Bay. Monomoy Island, which extends like a long finger to the south of Cape Cod, was once a spit connected to the Cape (Figure 6–10). Storms have broken through Monomoy, forming a tidal inlet at the north end and a barrier island to the south. The spits and barrier islands on Cape Cod are obvious examples of the spit theory of barrier island formation. The abundant sediment supply and rapid rate of erosion makes it possible to actually see the spits in the process of being formed.

Mid-Atlantic coast. There appears to be a basic similarity between portions of the mid-Atlantic coast from Cape Cod to Cape Fear. Fisher has drawn attention to four coastal compartments that link together to form a barrier island complex. The north end of each complex is terminated with a northern cuspate spit (Figure 6–11). The northern spit is wrapped around the end of the island and marks the point where sand enters deeper water. As the spit builds to the north, a submerged shoal area is laid down first, followed by the development of the spit on the submerged platform. Race Point on Cape Cod, Sandy Hook in New Jersey, Cape Henlopen in Delaware, and Cape Henry, Virginia, are examples of northern spits.

FIGURE 6–11. Coastal compartments on the Middle Atlantic coast include: I, northern spit; II, eroding headland; III, southern spit; and IV, barrier island chain.

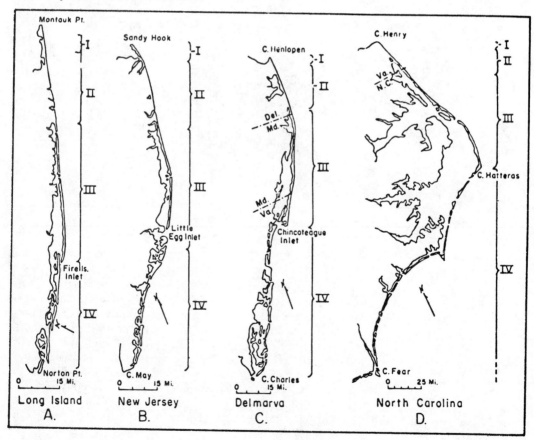

The middle compartment of the barrier island complex is the eroding headland. On Cape Cod and Long Island, where the spits are rapidly extending from the middle, the eroding headland is quite broad. Farther south on the Delmarva Peninsula (Delaware-Maryland-Virginia) and in North Carolina, the headland areas are considerably smaller (Figure 6–11). The broad headland area to the north can be accounted for by the large source of sediment where the coast is being cut into cliffs of glacial moraine and outwash. Farther south, sediment is supplied to the coast by an extensive network of rivers.

The southern spit develops to the south of the eroding headland. Generally, the southern spits are longer and better developed than the northern spits because the dominant longshore current direction is to the south. Nauset Beach and Monomoy Island on Cape Cod, Fire Island on the south shore of Long Island, and Assateague Island are examples of southern spits.

Beyond the southern spit, the barrier islands are far removed from their potential source area, and it is difficult to call on spits to account for the formation of the islands (Figure 6–12). For these islands, it may be more reasonable to use the ridge submergence hypothesis proposed by Hoyt. The barrier is-

FIGURE 6–12. Cedar Island, Virginia, on the Delmarva Peninsula, is a long, narrow barrier island backed by a broad salt marsh with tidal channels.
Photo by R. A. Davis.

lands along the southern extent of the Delmarva Peninsula formed as spits during a lower sea level stand, then migrated landward while sea level was rising. The eroding headlands provided the initial sediment for the barrier islands, but the islands advanced landward and incorporated the same body of sediment in the advance. As the islands moved landward, the submergence of the dune ridge played an important role in the continued presence of the island.

Outer banks of North Carolina. The Outer Banks along the coast of North Carolina provide beautiful examples of barrier islands (Figure 6–11). The Outer Banks extend south from Oregon Inlet, looping past Cape Hatteras and Cape Lookout, to Cape Fear. Pamlico Sound is a broad drowned river system preserved behind the Outer Banks. If the Outer Banks formed as spits at present sea level, mainland beaches would have been formed along the shore of Pamlico Sound before the spits had had sufficient time to close off the lagoon. A barrier ridge formed at the outer edge of banks during the Pleistocene and was modified by waves and currents since its formation. Portions of the earlier barrier are still preserved as shoals behind the Outer Banks.

The Wright Brothers made their first successful flight on the sand dunes along Nags Head. Cape Hatteras and Cape Lookout are being preserved in their natural state as part of the National Seashore. Extensive shoals exist south of the capes, where sand has accumulated due to converging longshore currents.

Along the North Carolina shore, borings were made into the soil horizons beneath the barrier islands to unravel the history of the islands. Where a soil horizon is present, the barrier is considered a primary barrier that developed by submergence of a coastal dune or ridge. Where the soil zone is not present in the borings, the island is considered of secondary origin, formed by the elongation and breaking off of spits.

Sea islands of South Carolina and Georgia. The sea islands of South Carolina and Georgia are distinctly different from the barrier islands to the north and south. In contrast to the long, slender barrier islands, the sea islands are short, stubby islands backed by broad marshes.

The sea islands have three distinct origins: erosion remnant islands, marsh islands, and beach ridge islands. The erosion remnant islands include Cumberland Island National Seashore and Sapelo Island. The erosion remnant islands are sand covered and irregular in outline. They formed where Pleistocene ridges are submerged beneath the sea to form mounds that project above sea level. The mounds are plastered with a new layer of sand on the seaward side, forming the present beaches and low dunes.

The marsh islands consist of various plants, mostly *Spartina patens*, the saltmeadow cordgrass, growing on clay, silt, and peat. The silt and clay are trapped by the stems and roots of the plants, which die and form peat. As sea level rises, the marsh island grows upward to keep pace with the flooding sea water. The marsh islands thrive in protected areas behind the main sea islands and do not survive on the exposed open coast.

The beach ridge islands are long, thin ridges or parallel strips of sand that protrude through the marsh. The beach ridge represents the active portion of the island where sand is being added to the front of the island. Ossabaw, Skidaway, and Hilton Head Islands are examples of beach ridge islands that have built seaward of erosion remnant islands.

Barrier islands, capes, and keys of Florida. Along the east coast of Florida, long, narrow barrier islands once again become the rule. Some of the barrier islands along the Florida coast are not separated from the mainland by a broad lagoon, but rather have a narrow lagoon parallel to the shore, which forms part of the intracoastal waterway. The islands that are fairly long are separated by widely spaced tidal inlets (Figure 6–5). In the northern part of Florida, the islands are covered with a mixture of fine quartz sand and shell fragments. Since most of the Florida peninsula is a domed layer of limestone, the quartz sand was carried to Florida from the north. The amount of quartz sand diminishes to the south along the coast, and is replaced by finely ground shell fragments. The southward movement of sand parallel to the Florida coast suggests that the barrier islands were formed by spit growth. Beautiful white sand beaches are found along the Gulf Islands National Seashore, on the northwest coast of Florida and Alabama.

The capes of Florida project seaward in a series of overlapping spits. For example, Cape Canaveral, on the east coast of Florida, and Cape Romano and Cape Sable, on the west coast, were formed by converging currents along the shore. The locations of the capes are controlled by the presence of extinct Pleistocene coral reefs on the shelf. At present, extensive reefs grow along the coast south of Miami. When sea level was lower, during the Pleistocene, the reefs grew farther seaward on the shallow continental shelf. Currents converged at the reefs, forming a gyre, which promoted the growth of the spits and maintained the positions of the capes.

The Florida Keys, a string of islands extending southwest from the tip of Florida, are a combination of coral reefs and mangrove islands. Where nutrients and a hard substrate are available, coral reefs have developed in the tropical climate. However, where the bottom is covered by soft sediment, mangrove trees become established and extend their roots into the bottom. The roots act as baffles and trap sediment to form an island. At Marquesas Key, about 35 kilometers west of Key West, the mangrove islands have formed a ring around a mud-bottomed lagoon that resembles an atoll.

Chenier plain of Louisiana. A chenier plain is present on the coast of Louisiana, to the west of the Mississippi Delta. A chenier is a long, sandy ridge, well above the high-water line, with fine-grained marshy sediment on its landward and seaward sides. Chenier comes from the French word *chêne,* meaning oak, for the large stands of oak that cover the tops of the islands. The sandy soil on top of the chenier provides an ideal environment for the oaks. The cheniers are between 50 and 500 meters wide and 10 to 50 kilometers long. However, the sand on the chenier usually only extends about 1 to 2 meters below sea level, and the height of the dunes is only 3 to 5 meters.

The cheniers were formed by the westward transport of sand, silt, and clay from the Mississippi Delta. The location of the mouth of the river shifts with time, and when it is close to the chenier plain, the amount of sand, silt, and clay supplied by the river is large. When the river shifts away from the chenier plain, deposition ceases and the sediment is reworked by the waves. During the reworking, the silt and clay are moved seaward and the sand is left to form a beach. When the river shifts toward the chenier again, a second phase of deposition starts and the chenier is isolated as a sand ridge within mud. With time, the muds are compacted to form lakes or freshwater marshes between the ridges. The cheniers differ from true barrier islands in that they do not migrate shoreward, but remain fixed within the surrounding mud.

On the east, the chenier plain is bordered by the Mississippi Delta, and to the west, it merges with the barrier islands of the Texas coast. Chenier plains are also present in the Rhone Delta in Germany, and in Surinam in northeastern South America.

Texas coast. Some of the best-developed barrier islands in North America are found along the Texas coast of the Gulf of Mexico. The major barriers include Padre, Mustang, Saint Joseph, Matagorda, and Galveston Islands. The barrier islands range in length from 25 kilometers for Saint Joseph Island to 180 kilometers for Padre Island, and rise up to 80 meters above sea level.

Most of the barrier islands on the Texas coast have broad, flat beaches and grass-covered dunes. On Padre Island National Seashore, wide aeolian or wind-blown flats extend behind the dunes. The flats, which are submerged only during storms, are covered by thin layers of algal mats.

Hurricanes often sweep across the Gulf of Mexico in late summer and early fall and spend their full fury on the Texas coast. The storm surges that accompany hurricanes force water across the barrier islands. Many of the homes on the barrier islands are built on stilts so that the storm surge will pass beneath the homes without destroying them. One of the most destructive hurricanes in American history occurred on Galveston Island in 1900. At that time, Galveston was a shipping center for the Texas coast, and also a resort area with a beautiful beach. The storm surge piled water up in Galveston Bay behind the barrier island and flooded the city from the back side. More than 6000 persons lost their lives on Galveston Island during the hurricane.

During the spring and summer, low waves from the Gulf of Mexico lap gently against the Texas beaches. During late fall and winter, sharp cold fronts, or northers, can cause a rapid drop in temperature with a sudden increase in wind. Their effects are usually short-lived and they do not account for much coastal erosion.

The barrier islands along the Texas coast are backed by a parallel set of Pleistocene barriers along the bay side of the lagoons. The inner set of barrier islands was formed at high sea level during the interglacial period before the last Wisconsin glacial advance. The Rio Grande, Brazos, and Colorado Rivers deposited deltas that were redistributed along the shore to form barrier islands. With the growth of the Wisconsin glacier, the shoreline retreated about 80 to 230 kilometers seaward of the present barrier island chain. The Texas rivers were clogged with sediment, which formed a new source for sand along the coast. With the recent rise in sea level, the new set of barrier islands developed along the coast. Sediment for the islands was supplied by longshore transport in the form of spits and by onshore transport of sand, which was deposited on the coastal plain.

Washington, Oregon, and California. Along the Pacific coast of North America, barrier islands are not common, but large spits and associated barrier islands are present. The west coast of North America is a typical collision coast with a relatively straight shoreline and parallel mountain ranges along the coast. In southern California, the coast has been broken up by a series of block faults. Some blocks have been dropped down to form deep basins, and other blocks have been lifted to form islands. Catalina Island and the islands off the coast of Santa Barbara are examples of fault block islands and would not be considered barrier islands. Farther north along the California coast, most of the shore is bounded by rugged cliffs, such as the Big Sur area. Although small spits form at the mouths of some of the bays, the large spits form barriers at Morro Bay and Humboldt Bay, California.

Beautiful spits are found associated with river mouths on the Oregon and Washington coasts. The spits along the Oregon coast have a completely different character from the barrier islands on the east coast (Figure 6–13). The high tidal range and

FIGURE 6–13. Nestucca Spit on the Oregon coast was built to the south by longshore currents. Large rip channels are present along the ocean beach.
Photo by W. T. Fox.

large wave energy produce spits with coarse sand and a steep foreshore. The sand is supplied to the coast by rivers that cut through the coast ranges. The rivers have a steep gradient and produce relatively coarse sands along the coast. The high waves result in large rip current systems that cut into the spits. Large spits are also present at Willapa Bay and Grays Harbor, north of the Columbia River in the state of Washington.

Alaska. Several large spits and barrier islands are found along the southern coast of Alaska. The islands of Alaska were formed by sediment carried to the sea from rivers draining large glaciers. The barrier islands on the Alaskan coast still lie in their pristine beauty, undisturbed by humans. They provide a thought-provoking contrast to the condominium-covered barrier islands in the southeastern United States.

Summary A barrier island is a single elongate sand ridge or multiple ridges, with vegetated dunes, which rise slightly above high-tide level. A spit is similar to a barrier island, but one end is attached to the mainland. Spits and barrier islands are separated from the mainland coast by a lagoon or salt marsh. Barrier islands are separated from each other by tidal inlets, which provide a passageway for ebb and flood currents.

Several different theories have been proposed for the origin of barrier islands. Elie deBeaumont suggested the emergence of submarine bars to account for barrier islands. Karl Gilbert proposed that barrier islands were formed from spits that were breached by waves during storms. W. D. McGee thought that barrier islands were formed by the drowning of a coastal ridge during a rise in sea level. Douglas Johnson, John Fisher, and John Hoyt later supported the theories of deBeaumont, Gilbert, and McGee, respectively. Recent evidence points toward a multiple origin of barrier islands.

Barrier islands migrate onshore, offshore, or along the coast. Onshore migration is caused by the movement of sand into the lagoon through tidal inlets or by overwash fans that pass through breaks in the dunes. The islands move offshore when additional sand is supplied by rivers. Alongshore movement is caused by strong longshore currents that build spits on the ends of the islands.

More than 12 percent of the world's coasts are bordered by barrier islands or spits. About 76 percent of the barrier islands are found along the trailing edges of spreading continents. About one-third of North America is bordered by barrier islands, which form an almost continuous string of islands along the southeastern Atlantic and Gulf of Mexico.

Selected Readings BRUUN, P. *Stability of Tidal Inlets, Theory and Engineering.* Amsterdam: Elsevier Scientific Publishing Company, 1978. An advanced text covering mathematical theory and engineering designs for tidal inlets between barrier islands.

COATES, D. (ed.). *Coastal Geomorphology.* Binghamton, N.Y.: Publications in Geomorphology, State University of New York, 1972. A collection of original papers on geology, ecology, and biology of barrier islands. Includes nontechnical papers that can be understood by the general public.

COX, W. S. *Oregon Estuaries.* Salem, Ore.: Division of State Lands, 1973. Maps, descriptions, and air photos of the spits and estuaries on the Oregon coast.

CRONIN, L. E. (ed.). *Estuarine Research, Volume II, Geology and Engineering.* New York: Academic Press, Inc., 1975. An excellent collection of research papers on geologic and engineering aspects of barrier islands and tidal inlets.

LEATHERMAN, S. P. (ed.). *Barrier Islands from Gulf of St. Lawrence to the Gulf of Mexico.* New York: Academic Press, 1979. A collection of short nonmathematical papers on the barrier islands along the east coast of North America.

LEATHERMAN, S. P. *Barrier Island Handbook.* Amherst, Mass.: University of Massachusetts, 1979. A basic introduction to barrier islands for the general public. Good illustrations and lists of plant species.

LEATHERMAN, S. P. (ed.). *Geologic Guide to Cape Cod National Seashore.* Amherst, Mass.: National Park Service, 1979. A field trip guidebook to geologic localities in the National Seashore, with a collection of general papers about the Cape.

SCHWARTZ, M. L. (ed.). *Barrier Islands, Benchmark Papers in Geology.* Stroudsburg, Penn.: Dowden, Hutchinson and Ross, Inc., 1973. A collection of early papers relating to the origin of barrier islands, which lead up to the current ideas on the subject.

SCHWARTZ, M. (ed.). *Spits and Bars, Benchmark Papers in Geology.* Stroudsburg, Penn.: Dowden, Hutchinson and Ross, Inc., 1972. A collection of geologic papers on the evolution and development of spits and offshore bars.

STRAHLER, A. N. *A Geologist's View of Cape Cod.* Garden City, N.Y.: Natural History Press, Doubleday and Company, 1966. An excellent introduction to the geology and coastal processes of Cape Cod, for the general public.

Common loon

eider

scoter

bay is cobalt blue

sand   snow

drift line debris

frozen back marsh

bayberry black cherry

fox tracks

snow bunting

rabbit

Plum Island, Newburyport, Mass.
1-28-82
2 pm   4" of fresh snow   28° blustery

page from sketchbook

CWL

Short-eared owl over low scrub in back dunes

SEVEN

Beach and dune ecology

Introduction to Beach and Dune Ecology

The ecology of barrier islands is closely intertwined with the sea. The sand and mud of the islands and marshes are supplied by the waves and currents and molded by the wind and vegetation. The climate of the barrier islands is milder than the adjacent mainland, warmer in the winter, and cooler in the summer. The spring blossoms arrive early on the islands, and the fall colors remain into late autumn. On the beach face, breaking waves and shifting sands produce an inhospitable environment for a good portion of the year. However, beaches, dunes, salt marshes, and tidal flats provide abundant habitats for a diverse flora and fauna.

Hardy grasses on the barrier islands survive and actually thrive on burial by sand. The grasses trap sand, which builds up to form dunes. Many of the plants on the dunes are especially adapted to salt spray and have evolved mechanisms for extruding salt from their tissues. Seeds and roots carried by the waves are washed up along the beach, forming drift lines. Hardy pioneer species of plants become rooted in the drift lines and start the formation of dunes. Sand that blows off the beach is trapped by the plants to form large coastal dunes.

Many small invertebrates, including crabs, clams, beach hoppers, and pill bugs, live on and in the beach. Some of the animals move back and forth with the waves, while others extend their arms or tentacles into the waves to capture food. Small and large mammals live in the dunes and maritime forests, and venture onto the beach at night to feed on the feast brought up by the waves and left by the falling tide.

Several different types of shorebirds live along the beach and fish in the surf zone. The gulls and terns dart and soar along the beach, while the small sandpipers scamper back and forth with the waves.

The art of field sketching

Simple methods of field sketching outdoors can be most helpful for many aspects of coastal study. One does not need an advanced skill in drawing but, rather, an ability to record accurately what is being identified, whether it be the height of a beach scarp, the shape of a wave pattern, a particular rock formation, or a variety of plants and animals seen in a certain area.

Keeping written as well as drawn field notes helps the student learn and record key characteristics, which can then be redrawn in detail on return indoors (Figure 7–1). All field naturalists have at one time or another kept records of this type. It is much easier and more accurate to refer to sketched notes, no matter how crude, than to rely on memory as to whether the scarp was four feet high or two, whether the bird seen was white on the head, and so on. A few quick, diagnostic lines are enough to give the overall sense of a form. Less important features can be added later. (See the opening drawing for this chapter.) Numerous illustrations in this book were done from previous field sketches.

FIGURE 7–1. Clare Walker Leslie sketching marsh grass in a
salt marsh at Plum Island, Massachusetts.
Photo by W. T. Fox.

Field sketching materials should be kept simple. A small,
ringed sketch pad; a pencil or felt-tipped pen; perhaps some colored pencils; a pair of binoculars and, or a magnifying glass; a
6-inch ruler; a few plastic bags or bottles for collecting; and a
field guide book or two all carried in a backpack are enough for
a day's expedition. For further information on the art of field
sketching, refer to Clare Walker Leslie's *Nature Drawing: A Tool
for Learning,* Prentice-Hall, Inc., 1980.

Beach and
Barrier Island
Habitats

The habitats across a spit or barrier island are shown in a dia-
grammatic cross-section (Figure 7–2). The surf zone or sand
flats extend along the ocean side of the island. Waves supply
food for small animals living in the sand on the beach, and
wash plants and seeds to the drift line at the back of the beach.

The primary dune, covered with beach grass, traps sand
blown off the beach. A trough or swale often forms behind the
primary dune. Secondary dunes covered with grasses and perennial plants become established behind the primary dune. Annual plants also grow on the open spaces between the clumps
of grass. Small mammals that are common on the mainland
also thrive on the dunes.

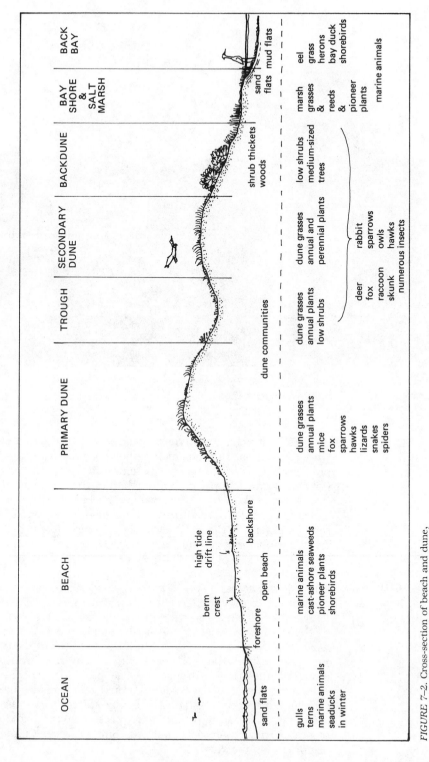

FIGURE 7–2. Cross-section of beach and dune, and related ecology.
Drawing by C. W. L.

Scrub thickets and medium-sized woods grow in the protected backdune area (Figure 7–2). On some islands, maritime forests grow, using salt from the sea for nutrients.

On the bay shore, marsh grasses and reeds grow along the edge of the island. The salt marshes, flooded by the tides twice each day, give way to intertidal mud flats. Wading and floating birds fish along the bay shore of the marsh.

Beach ecology

Sandy beaches, which are often thought of as prime recreational areas, also bear the full brunt of ocean waves and are subjected to shifting currents with each storm. Plants have a difficult time becoming established on the beach because they are carried away by the waves and tides. Seeds and plant fragments are laid down on drift lines marking the farthest advance of the waves. The highest drift lines provide a stable habitat for the seeds to take root and form small tufts of grass, but lower drift lines are reworked by the waves so that plants don't become firmly established in the sand. Some of the plants on the beach are removed by winter storms and therefore only occupy the beach during the summer. Embryo dunes may form in the higher drift lines where blades of grass trap sand around their bases.

Animals that live on the beach are adapted to burrowing into the sand so they can escape the waves, or they are fast crawlers who scurry back and forth with each wave. Various types of sturdy clams and worms live on the foreshore and depend on rapid burrowing to remain beneath the pounding surf. Pill bugs, beach hoppers, and crabs scurry back and forth on the beach and forage in the wave swash for food.

Swash zone inhabitants. On the east coast of North America, most clams are found on the protected sand flats behind barrier islands. On the Pacific coast, some of the most delicious clams are collected in the surf zone on the open coast. Ricketts, Calvin, and Hedgpeth have described the beach organisms in their delightful and informative book, *Between Pacific Tides.* The razor clam, pismo clam, and bean clam are the three most common surf clams on the Pacific coast.

The Pacific razor clam, *Siliqua,* lives in the foreshore along the sandy beaches of Washington, near the mouth of the Columbia River. The adult razor clam has a shell about 15 centimeters long and 5 centimeters across, with fragile, shiny valves (Figure 7–3). It is quite distinct from the Atlantic razor clam, *Ensis,* which lives in the bays on the east coast, and is long and narrow with an oval cross-section. The foot of the Pacific razor clam can extend half the shell length and is an effective burrowing tool. The pointed tip, which is used for digging, expands to form an anchor that holds fast when the clam retracts into the sand. When a razor clam is laid flat on its back on the moist sand, it takes only a few seconds to bury itself completely. Up to 80 percent of Pacific razor clams have a commensal nemertean worm,

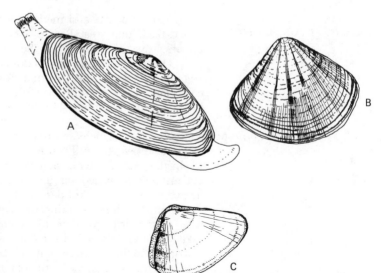

FIGURE 7–3.
Pacific coast clams: (A) Pacific razor clam, *Siliqua* (length: 15 cm); (B) Pismo clam, *Tivela* (length: 10 cm); (C) little bean clam, *Donax* (length: 3 cm).
Drawing by C. W. L.

Malacobdella, which is attached to the clam's gills by a sucker and feeds on plankton brought in by the siphon. The word commensal means companion at meals, and the worm shares its feast with the clam without adverse effect on either organism.

On the California coast, the pismo clam, *Tivela,* takes the place of the razor clam in the surf zone. The adult pismo clam is 10 to 15 centimeters across and is sought after as a delicacy in southern California (Figure 7–3). Although it was harvested commercially in the past, its numbers have significantly decreased in recent years. It was once abundant on the foreshore, but it is now found in waist-deep, breaking waves. The pismo clam lives in the surf zone where the constantly agitated waves aerate the water, providing abundant oxygen. When the clams are moved to the protected tide flats where the temperature, salinity, and tide range are similar to open coast, the clams die within a few days. Pismo clams live buried in the sand with their hinge lines facing toward the surf. They are not as fast burrowers as the razor clams, but they eject water from their mantle cavity to give them a jet assist in burrowing. Pismo clams have light and dark bands on their shells, which can be used to tell their age. The light bands form in the summer and the dark bands in the winter. It takes about 4 to 5 years for a pismo clam to reach the legal size of 11.5 centimeters (4½ inches). Many of the clams are 15 to 20 years old, and the oldest recorded clam had reached the ripe old age of 53 years.

The bean clam, *Donax,* is a small triangular-shaped clam that lives in clusters on the California coast. It has a smooth, thick shell that withstands the pressure of breaking waves (Figure 7–3). Bean clams live just below the surface of the sand, where the uprush of the waves provides a constant supply of food. A small hydroid, *Clytia,* grows on the bean clam shell and extends above the sand surface, giving away the hiding place of

the clam. On the California coast, the bean clam does not move up and down the foreshore with the tide, which is the practice of other species of *Donax* on the coasts of Texas and Japan.

The bloodworm, *Euzonus,* a small, bright red worm, also lives in the Pacific coast surf zone. The bloodworm is 4 to 6 centimeters long and burrows a slender tube in the sand where it is damp but not mushy (Figure 7–4). The bloodworms occur in patches along the coast, and where they are present, they reach abundances of 25,000 to 30,000 per square meter. McConnaughey and Fox at Scripps Institute of Oceanography estimated that a worm-infested beach, a kilometer long, would contain 95 million worms or 4 tons of fresh red bloodworms.

The mole or sand crab, *Emerita*, lives in the swash zone, filtering food from the wave backwash (Figure 7–5). The mole crab is a fascinating little creature, 3 to 5 centimeters long, which lives buried on the top layer of sand in an egg-shaped shell. Unlike other crabs that move equally well in all directions, the mole crab always moves backward. When feeding, the mole crab stands on end, head up and facing into the surf. As the wave rushes up the beach, it hunkers down into the sand with just its eyes and first pair of antennae above the surface. With the backwash of the wave, the mole crab stands up and extends its second pair of large antennae in the form of a *V* to filter planktonic organisms out of the waves. The large antennae are pulled through bottlebrush-like appendages to remove the small fragments of food. The mole crab extends its antennae several times during each wave. Mole crabs live in dense colonies, occupying almost all of the available space. The colony moves up and down the beach in unison to stay in the zone of breaking waves. Although they are most abundant in California, they are also found along the east coast of North America and extend in the Pacific from Alaska to Patagonia.

Beach organisms. The beach is inhabited by beach hoppers, crabs, and pill bugs, which venture into the swash zone in search of food. The pill bugs are small isopods, *Eurydice,* on the east coast, and *Alloniscus* or *Tylos* on the Pacific coast. Pill bugs burrow just beneath the sand surface, above the high-tide line, living a molelike existence. The isopod, *Tylos,* is restricted to the upper beach on the southern California coast (Figure 7–6). Small isopods, less than a millimeter long, also live in the pore spaces between the sand grains.

At dusk and in the early evening, the beach hoppers or amphipods, *Talitrus* and *Orchestoidea,* scavenge among the debris left on the beach. *Orchestoidea* is a dapper animal with an old-ivory-colored body and long, bright orange antennae (Figure 7–6). The hoppers, which are up to 6 centimeters long, avoid getting wet in the waves, but keep their bodies moist during the day by burrowing beneath the sand.

On tropical beaches, the scavenger role of the beach hoppers is played by the ghost crab, *Ocypode* (Figure 7–7). The

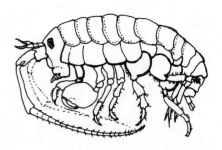

FIGURE 7–6.
Pill bug or isopod, *Tylos* (length: 1.5 cm) and beach hopper, *Orchestoidea* (length: 5 cm).
Drawing by C. W. L.

ghost crab is a nocturnal scavenger that can still be seen roaming the beach in the early dawn. The ghost crab spends the heat of the day in its burrow and returns to the water only to wet its gills. Along the southeast coast of the United States, ghost crabs have been used as an indicator of beach use. With heavy pedestrian use, the number of ghost crabs has actually increased, probably because they are living on food scraps scattered by bathers. Where vehicle traffic is heavy on the beach, the ghost crab population is diminished, with all the crabs quite small or young.

Drift line animals. The Portuguese man-of-war, *Physalia*, lives in the open sea, but often washes up on the beach. The Portuguese man-of-war drifts with the wind and warm currents with its purple body and clear sail extending above the water. The man-of-war is a siphonophore, a colonial organism that can deliver a painful sting from its loose dangling tentacles. The tentacles with up to 750 nematocysts, or stinging cells, extend up to 20 meters beneath large specimens. The tentacles retain their stinging potential even when they are washed up on a beach. With an extended interval of onshore winds, some Florida and Gulf coast beaches are littered with thousands of Portuguese man-of-wars.

On the Pacific coast, a small floating organism looks like a miniature version of the Portuguese man-of-war, but is an entirely different animal. When it washes up on the beach, it looks

FIGURE 7–7.
Ghost crab *Ocypode* is a scavenger on the upper beach (length: 50 cm).
Drawing by C. W. L.

184

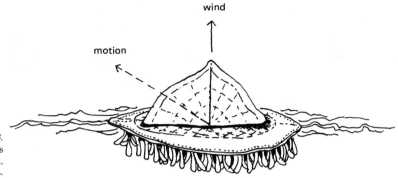

FIGURE 7–8.
Velella. "By the wind sailor" drifts
with winds and currents.
Drawing by C. W. L.

like the discarded cellophane wrapper from a cigarette package. The Pacific form, *Velella,* is a chondrophore, which is a highly modified individual hydroid, about 5 to 10 centimeters across, that has taken up life on the high seas. The animal is blue to purple below a transparent sail. The small, triangular-shaped sail cuts a diagonal across the long axis of the body (Figure 7–8). The specimens that drift ashore on the Pacific coast of North America have the sail set askew along the long axis extending from the northwest to the southeast. In light, southerly winds, *Velella* tacks to the left, which would force it offshore. With strong winds, *Velella* spins around and follows the wind more closely. The animal appears on the beach in the spring after the first strong southerly or westerly wind. In the Southern Hemisphere, *Velella* that reach shore have the sail set from northeast to southwest. It is thought that there are equal numbers of *Velella* with different sail settings in the tropics. When the wind blows, they are separated, with some heading north to California and others heading south to Peru and Chile.

Beach mating. Although not permanent residents of the beach, several marine species move on the beach to mate or lay eggs. Large sea turtles, including the loggerhead turtle, come onshore to lay small leathery eggs in the sand. When the eggs hatch, tiny turtles scurry to the sea. Many are lost as they cross the sand or struggle in the surf. Enough turtles survive to maintain the species, but their numbers have been greatly reduced in the Caribbean, where their shells have been collected for making combs and jewelry.

The grunion, *Leuresthes,* a smeltlike fish about 15 centimeters long, provides many residents of southern California with a holiday at the beach. On the second, third, and fourth nights after the full moon in March, April, May, and June, the grunion swarm on the beaches to mate. During the high spring tide, the grunion swim up the beach with the breaking waves. They come in pairs, with the female digging into the sand, tail first, and laying her eggs. The male arches around her and fertilizes the eggs (Figure 7–9). Then they slip back into the sea on the next wave, if they are not caught by hungry Californians.

A.

B.

FIGURE 7–9.
(A) Female grunion has just deposited eggs (tailfirst) into wet sand. Male partner soaks sand with milt, fertilizing the eggs (length: about 15 cm). (B) Egg laying completed, the female leaves her nest to return to the sea.
Drawing by C. W. L.

During the following new moon spring tide, which is higher than the full moon spring tide, waves wash over the eggs, which hatch and enter the sea. The timing of the grunion runs is critical to their survival. If the eggs were deposited on the new moon spring tide, they would be stranded on the beach when they hatched 15 days later. If they were laid on the neap tide, they would be washed away on the next high tide.

Dune ecology   Coastal dunes are constructed by sand blown from the beaches and exposed tide flats, and held in place by dune grasses. Along the west-facing coasts of Europe and North America, the steady westerly winds provide a constant supply of sand to the dunes. On the east coast of North America, the onshore winds are provided by the afternoon sea breezes, but storms with strong east winds supply most of the dune sand.

Pioneer species.   Dune plants play an important role in the formation of dunes, and different plants are prominent at different stages in the life cycle of a dune. The first stage is the development of an embryo or foredune, which is colonized by pioneer species that are very resistant to salt (Figure 7–10). The formation of the embryo dune begins on the drift line where seeds are brought ashore among the flotsam and jetsam. The decaying organic matter provides nutrients and moisture for the germination and growth of the seeds. Although several drift lines can often be observed on a beach, plants are generally re-

186

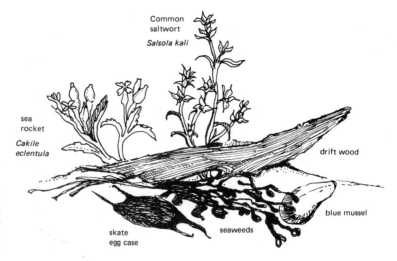

FIGURE 7–10.
Embryo dune begins at the drift
line where flotsam and jetsam
collect with pioneer plants to help
secure and build up sand.
Drawing by C. W. L.

stricted to the highest drift line, which marks the farthest reach of storm waves during spring tides. Lower drift lines are the product of subsequent storms that sweep the beach clean of young plants.

The plants that prosper in the drift line are considered pioneer species. Common pioneer species on the east coast include the sea rocket, *Cakile maritima*; saltwort, *Salsola*; and shore orache, *Atriplex*. These species occur on the coasts of Norway, Labrador, and New Foundland in the north, down to Portugal and Florida in the south. The sea sandwort, *Honkenya*, and sand couch grass, *Agropyron*, are also found with the drift-line species (Figure 7–10). On the Pacific coast, the sea rocket, *Cakile edentual*, is one of the first colonizers each year on the seaward edge of the dunes.

In Europe, *Agropyron* is the dominant pioneer species in the embryo dune. It is tolerant to salt water, where it thrives on 1.5 percent salt in the soil and can persist with up to 6 percent salt. *Agropyron* is also an important pioneer species because of its extensive root system. The tillers, or shoots, reach into the wind to trap the sand while the roots bind the sand that has been deposited. The earliest tillers lie flat along the ground and give rise to a radiating pattern, or rosette of shoots, depressed in the sand. The underground rhizomes extend about 5 to 30 centimeters from the single primary root and supply a new group of tillers to trap the sand. The rhizomes grow laterally during the summer, with the tips turning upward in the fall to form new tillers the following spring. *Agropyron* are limited to about 30 centimeters of burial per year, but have almost unlimited horizontal rhizome growth. Therefore, the initial embryo dunes are low mounds that provide a base for the more active beach grasses.

Many other plant species are brought in as seeds in the drift line and become established briefly on the beach. Other

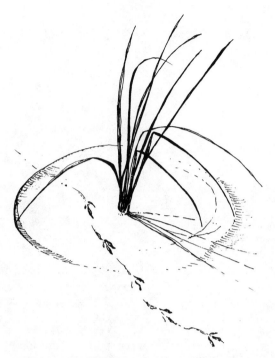

FIGURE 7–11. American beach grass with wind ring
and crow tracks in the sand.
Drawing by C. W. L.

species, such as morning glory, *Ipomoea,* trail off the dunes as
vines and extend onto the backshore. *Ipomoea pes-caprae,* the
railroad vine, provides beautiful flowers on the otherwise desolate beaches of the southeastern United States.

Beach and dune grasses. When the embryo or foredune
has become established, various species of beach grass take over
and the dunes rapidly increase in height. In North America, the
most abundant species is American beach grass, *Ammophilia
breviligulata,* which occurs on the east coast, north of Isle of
Palms, along the South Carolina shore, and on the shores of the
Great Lakes (Figure 7–11). In southeastern United States, sea
oats, *Uniola,* extend southward from Cape Henry, Virginia. The
sea oats have long stems with a plume of seeds on the top (Figure 7–12). The sea oats, which once were collected for dried
flower arrangements, are now protected in many states.

On the Pacific coast of North America, various perennial
grasses, including dune grass, *Elymus,* and the seashore blue
grass, *Poa,* form the dune. The European beach grass, or mar-
ram grass, *Ammophilia arenaria,* was introduced to the Pacific
coast in the late 1800s, in a program of dune stabilization. The
hardy grass grew up and down the coast and is now one of the
most common beach grasses.

FIGURE 7–12. The Florida beaches are typified by the presence of tall sea oats, *Uniola*; terns; and low, rambling beach bean, *Lathyrus*.
Drawing by C. W. L.

Ammophilia is a hardy strain of beach grass that requires burial by sand for continued growth. Beach grass has the ability to send out both lateral and vertical rhizomes so that it can survive up to a meter of burial each year (Figure 7–13). In an experiment conducted on Cape Cod, the ends were removed from a 50-gallon steel drum, which was placed over a clump of beach grass. The drum was filled with sand in the late spring and left on the dune. By the end of the summer, rhizomes had grown vertically through the drum and tillers of beach grass were sprouting on the top. Although *Ammophilia* grows rapidly when it is buried by sand, it will lie flat and disintegrate when dune growth ceases. Therefore, when the dune has reached its full height, the beach grass is often replaced by shrubs and thickets. Following a blowout, the beach grass is rejuvenated while the dune builds again to its former height.

The wind velocities drop significantly when the wind blows through the grass. Therefore, the sand that is being carried close to the ground settles around the blades of the grass. The beach grass has a high silica content and the roots do not

189

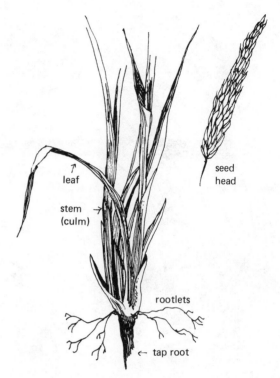

FIGURE 7–13. Close-up of American Beach grass, *Ammophila*. Drawing by C. W. L.

break down rapidly when they are buried. The interconnected network of roots holds the dune in a near-vertical face when the dune is being eroded. As the dunes grow higher and wider, they eventually form a single continuous dune ridge, which extends parallel to the shoreline.

American beach grass is more hardy than its southern counterpart, sea oats, and allows for higher dunes in the northeast than in the southeast. Beach grass begins to decline in vigor and is more diseased in Virginia and North Carolina. Sea oats do not form a dense cover, as beach grass does, but grow more slowly and form distinct clumps. Therefore, sea oats are not as effective as sand traps and produce the low, irregular dunes more typical of the southeast.

On the shores of the Great Lakes, *Ammophilia* thrives, forming some of the highest coastal dunes in the world. Along the ocean beaches, beach grass is somewhat restricted by direct contact with salt water, but it is resistant to salt spray. It is necessary for pioneer species to form embryo dunes before beach grass can become established. With the fresh water, beach grass is present in the embryo dunes and continues to cover the dunes as they build in height. Extensive dunes are present on the southeastern shore of Lake Michigan, where the dunes reach more than 70 meters in height. At Sleeping Bear sand dunes along the northeastern shore of Lake Michigan, the dunes reach more than 200 meters above the lake level. The dunes are slowly moving to the east, covering a growing forest, while an ancient forest is emerging from the trailing edge of the dune.

Dune scrub and thickets.   When dunes are fairly well established with a good cover of beach grass, other plants colonize the dune with wind-blown seeds. In the northern states where American beach grass is well developed, the dunes reach 10 to 20 meters above sea level. A wide variety of perennials, including dusty miller, *Artemisia,* beach heather, *Hudsonia,* and poison ivy, *Rhus radican,* fill in the spaces between the stands of beach grass (Figure 7–14). Plum Island, one of the northernmost barrier islands, on the coast of Massachusetts was named for the beach plum, *Prunus maritima,* which grows on the dunes in the shrub thicket.

It has been suggested by some hardy conservationists that poison ivy be planted along with beach grass to preserve the dunes from overuse. Although it may be effective against foot traffic, it still does not affect rubber tires.

FIGURE 7–14. Annual plants of the northeastern dunes. *Drawing by C. W. L.*

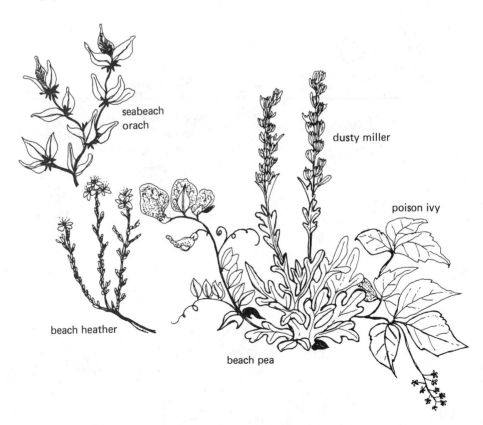

seabeach orach

dusty miller

poison ivy

beach heather

beach pea

Many of the species that are adapted to life on the dune are nitrogen-fixing plants such as bayberry, *Myrica*, and beach pea, *Lathyrus*. The nitrogen-fixing plants add nitrates to the sand, which helps support other plants and eventually establishes a soil for scrub thickets and forest.

On the southeastern coast, *Smilax*, a woody evergreen shrub, provides the greatest coverage in the low areas between the dunes. An herb species with conspicuous yellow ray-flowers, *Coreopsis*, is also abundant on the lee slopes and in the protected swales, indicating that it is tolerant of salt spray but cannot be subjected to sand burial.

The scrub thicket develops in the low area between dunes and the protected area behind the dunes (Figure 7–15). The scrub thickets include Beach plum, Bayberry—*Myrica*; Shadbrush—*Amelanchier*; three varieties of cherry, Chokecherry—*Prunus virginiana*, Fire cherry—*P. pensylvanica*, and Black cherry—*P. serotina*; Beach heather—*Hudsonia*; Chokeberry—*Aronia*; Briars—*Smilax*; Roses—*Rosa rugosa* and *R. virginiana*; and Bearberry—*Arctosaphylos*.

FIGURE 7–15. Scrub thickets between dunes.
Drawing by C. W. L.

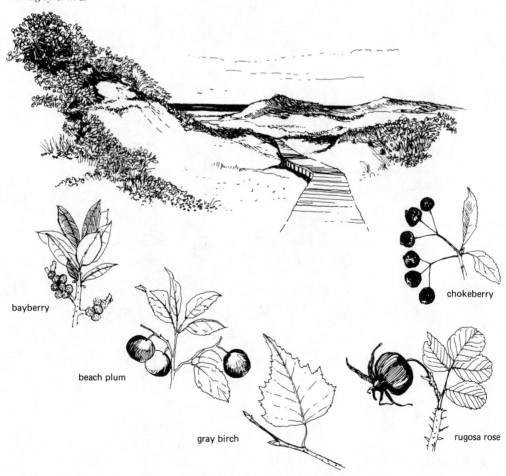

chokeberry

bayberry

beach plum

gray birch

rugosa rose

Barrier flats. The area between the dunes and the marsh is called the barrier flats. Where the dune ridge is poorly developed, the dunes do not provide much protection for the barrier flats, and shrub thickets grow between patches of grass. In the southeast, low dunes are subjected to frequent overwash during storms, and washover fans bury portions of the barrier flats. On the barrier islands along the Texas coast, broad barrier flats, which are called aeolian flats, extend behind the low dunes. The area is subjected to frequent overwash and flooding from the bay side, which leaves water ponded on the aeolian flats. In the hot Texas climate, evaporation of the salt water leaves a hard pan on the surface of the flats. A dark organic layer is preserved beneath the hard pan.

In the northeast where the primary dune ridge is 6 to 10 meters high, overwash is concentrated in certain areas along the dunes. The low breaks in the dunes are flooded when large storms coincide with high spring tides. Therefore, overwash fans are limited in extent and barrier flats are much smaller in the northeast.

Maritime forests. In the area between overwash fans, fairly well protected forests grow behind the dunes. The forest usually starts on the backside of the dune near the marsh, where water and nutrients are available. As the forest grows, it advances seaward behind the protective barrier of the dunes. The maritime forest needs to be blocked from the onshore winds that carry destructive salt spray across the dunes.

The Sunken Forest at Fire Island National Seashore, off the southern coast of Long Island, New York, provides an excellent example of a maritime forest. The forest includes American holly, pitch pine, red maples, black tupelos, and sassafras trees. The American holly in the forest is up to 30 centimeters (1 foot) across the base of the trunk and more than 150 years old. The large holly trees are often shaped by the salt spray carried by the wind (Figure 7–16).

The maritime forests play an important role in the barrier island ecosystem. Where primary dunes are well established, such as Plum Island, Massachusetts, and Fire Island, New York, the maritime forest covers the barrier flat from the back of the dune to the marsh. Where dunes are low, a thicket or grassland may grow behind the dunes and the forest is set well back on the island. On the Sea Islands of Georgia, a marsh develops behind the low dunes and the forest forms on old dunes or islands behind the marsh.

Beach ridges and ponds. As a spit develops, a series of curved beach ridges sweep around the end of the spit. The succession of dune development is nicely displayed on the accreting spit. A new deposit of sand is laid down as a submerged spit that finally builds up above mean high tide and forms a beach. The drift line at the back of the beach is cut off from the

FIGURE 7–16. American holly shaped by the coastal winds in the Sunken Forest, Fire Island National Seashore, New York. *Photo by H. Art.*

attack of waves as the spit continues to grow down the shore. The drift line provides the nucleus for dune development. The dunes increase in size and become stabilized with a cover of beach grass. Salt marsh will often form in the low slacks or swales between the dunes. As the spit increases in size, a debris line forms along the marsh side, and the marsh is eventually cut off from the salt water. The marsh also grows upward, with the area of low marsh being replaced by high marsh and eventually forest. When the salt water marsh is cut off from the tides, the rains flood the marsh, forming a freshwater marsh. With poor drainage, freshwater marsh becomes a bog and inhibits the forest growth, which moves up on the dune ridge.

The freshwater wetlands on the dunes provide resting areas and feeding stations for migrating birds. Canada geese and ducks make stops along the barrier islands off the New England coast on their flight south to the feeding and wintering areas along the southeastern marshes. Freshwater ponds are formed in the dunes where sand is removed by blowouts. Where the vegetation cover is removed from the surface of the dune, the sand will be blown out, forming a bowl-shaped depression. During periods of heavy rain, the water table on the barrier island rises and the blowout becomes a pond. Where the pond is standing for a considerable time, mud or clay may collect in the pond, sealing off the bottom of the pond and forming a perched water table. In areas of steady year-round rainfall, the dune ponds become permanent features on the landscape.

Where a small coastal lagoon is closed off in the process of dune development, a freshwater pond will form. The freshwater on a barrier island exists as a lens that is bouyed up above the salt water. The salt water is more dense and has a specific weight of about 1.025 times as great as fresh water. Therefore, the lens of fresh water extends down 40 meters for each meter above sea level. A pond that is 2 meters above sea level is the surface expression of a lens of fresh water between the pores of the sand that extends 80 meters beneath sea level. If the water on the barrier island is pumped from shallow wells, the water table will drop and the boundary between fresh and salt water will rise. This is known as salt water intrusion and can quickly ruin freshwater wells on a barrier island.

The vegetation found in the freshwater environments include rushes, sedges, cattails, and other marsh plants. The small ponds that develop in the dune swales and slacks do not have outflowing streams. Therefore, the ponds become acidic and form reducing environments with low oxygen content. Heath and sphagum moss communities similar to those found in bogs occur in the dune ponds.

Barrier island animals. Small animals that are common on the grasslands and forests of the mainland also inhabit the barrier islands and spits. Red foxes are often seen on the islands, and their tracks are common in the dunes. Raccoons nest in the trees of the maritime forest and forage along the bay shore for crabs and clams. Wildlife people claim that is it possible to tell fox scat from that of raccoons because fox scat contains bird feathers, while raccoon scat has broken-shell fragments. This may be based somewhat on the bad reputation that the fox has earned as a bird hunter, indicating that all feather scat is assigned to foxes. On several barrier islands, foxes are hunted to protect the bird populations.

Herds of deer are also present on some of the barrier islands, where they feed on small trees and shrubs. Their distinctive cloven-hoof print can be found in the dunes and, in the early mornings, on the beaches. The deer can easily swim from the mainland to the islands or from one island to another. The common white-tail deer is found on many islands, but the Sika deer, a native of Japan, is also present at Chincoteague National Wildlife Refuge in Virginia. The Sika deer was introduced to Chincoteague along with the wild ponies, which have become a local tourist attraction.

Tracks of many small animals can also be seen between the grasses and plants of the dunes. Rabbits, mice, and other small mammals scamper through the dunes in search of food. Hawks and owls hunt small mammals on the islands, the hawks by day and the owls by night.

Shorebirds   It is fascinating to watch the different types of shorebirds that congregate along the beach. Gulls and terns are aerial searching birds that spot their prey as they drift with the wind currents. Sandpipers and plovers are surface-searching and shallow-probing shorebirds that walk along the beach, and advance and retreat with the waves. The deep-probing shorebirds, wading birds, and diving birds, which spend most of their time in the salt marshes and tidal flats, will be discussed in Chapter 8. The large birds of prey roam the coast but nest in the marshes, so they will be considered with the marsh birds.

The shorebirds described in this chapter are but a few of the many birds found along the beaches, marshes, tidal flats, and rocky shores. Much of the information on the descriptions of the birds and their voices was derived from the *Peterson Field Guide Series,* published by Houghton Mifflin Company. The eastern birds are covered in *A Field Guide to the Birds East of the Rockies.* The Pacific coastal birds are included in *A Field Guide to Western Birds.* The birds of the Gulf of Mexico are best covered in *A Field Guide to Texas and Adjacent States.* Most of the birds covered in this chapter are common to both the Atlantic and Pacific coasts. Most of the birds of the Gulf of Mexico are included with the eastern birds. It is noted where birds are limited to the east or west coast of North America. Both coasts are discussed in the Golden Press book, *Birds of North America,* by Chandler Robbins. Since most of the birds are well known by their common names, the Latin names have not been introduced here.

Aerial searching birds   The aerial searching birds, which include gulls, terns, and skimmers, fly over the surf in search of prey. The gulls and terns demand attention with their antic displays and raucous calls (Figure 7–17).

Gulls are long-winged swimming birds with excellent flight. They are larger, longer winged, and longer legged than terns, and have square or rounded tails. The two common gulls on the east coast are the common and black-winged gulls. Easily adapted to the human environment, these two gulls are seen around inland trash dumps and along busy fishing wharves.

The laughing gull, which spends its winter from Florida to Venezuela, can be found north to Nova Scotia during the summer. It is a small gull with a dark covering on the wings that extends into black wing tips. In summer, it has a black head, which fades to a white head with a dark smudge in winter. It has an unmistakable strident laugh: "ha–ha–ha–ha–haah–haah–haah."

The glaucous gull is uncommonly found in the northeast, but it is one of the most common gulls on the Pacific coast. It is a large, chalky-white gull with frosty wing tips and a pale-gray mantle.

The terns are smaller, sleaker, and more streamlined birds than the gulls. Their tails are usually forked and their sharp,

laughing gull

osprey

Canada goose

common tern

double-crested cormorant

brown pelican

common loon

black skimmer

bufflehead

snowy egret

long-billed dowitcher

whimbrel

greater yellowlegs

black-bellied plover

sanderling

FIGURE 7–17. Birds of coastal beaches and marshes.
Drawing by C. W. L.

pointed bills are tilted downward. The terns are white or gray with black caps. The terns hover and plunge headfirst into the water for fish, but they do not swim like gulls. The gulls, who are their most common enemies, rob their nests of eggs and plague them to drop their fish.

The terns are cosmopolitan, but they migrate north and south with the seasons. Along the Atlantic coast, the least tern and common tern are found in the northeast during the summer, but they fly as far south as the Southern Hemisphere in the winter. The Forester's tern is found along the mid-Atlantic coast in the winter, and heads for the Arctic in the summer. The Arctic tern has one of the longest migratory flights, nesting in the high Arctic and wintering far down in South America. They stop off briefly along the east coast of the United States during their fall and spring migrations. The Sooty terns, which nest in the spring in the Dry Tortugas off the Florida Keys, spend up to 18 months in flight. They are occasionally seen farther north, where they are blown by hurricanes. The common and least tern are also present along the Pacific coast, where they breed in the Pacific northwest and migrate to South America in the winter.

The black skimmer can often be seen flying low above the surface of the water. The skimmer is black above and white below, with a bright red scissorlike bill. The skimmer flies low and dips its knifelike lower jaw into the water to catch fish. It is found from Cape Cod to South America and spends its winters off the southeast coast of the United States.

The brown pelican is a huge, dark, water bird with a wingspread of up to 2 meters (6½ feet). Its flight is unique, a few flaps then glide, with its large head back on its shoulders and its bill resting on its breast (Figure 7–17). When feeding, the pelican takes a few deep sweeps of the wing, lifting itself up, then plunges, bill first, into the water. Lines of brown pelicans often fly in close formation along the surface of the water, then rise to take their plunge. The white pelican, which lives in the west and central North America but winters in the southeast, does not plunge like the brown pelican, but scoops up fish while swimming. A few years ago, the pelicans were on the decline because of D.D.T. D.D.T. was spread on fields and concentrated in the water of the Mississippi and other rivers. It worked its way up the food chain and was finally concentrated in the shell of the pelican egg. Eggs were too thin and broke in the nest before the young could hatch. With the end of D.D.T. usage in the United States, the pelicans have made a strong comeback.

Probing and surface-searching shorebirds

The shallow-probing and surface-searching shorebirds have a wide diversity. They have specialized bills, so they can share a habitat with similar birds but exploit different sources of food. The sandpipers and plovers feed by sight, preying on surface fauna, insects, and pill bugs. Most other species feed by touch, with sensitive nerves in their bills that allow them to detect polycheat worms and other fauna living within the mud or sand. The oyster-catchers are large birds with long, chisel-tipped red bills for breaking open oyster shells. They also feed on other mollusks and crabs. Rails are compact, hen-shaped marsh birds that hide in the salt marsh and go after crabs at low tide. The shallow-probing and surface-searching shorebirds migrate to the Arctic in the summer, and winter in Central and South America. They are most abundant along the North American coast during the spring migration, which lasts from 1 to 2 months. They are less abundant during the fall because the migration extends over a 5-month period.

Although it is not possible to describe all the shorebirds, it is useful to be familiar with a few of the more common birds one might encounter along the shore. The Sanderling is the small, plump, active sandpiper of the outer beaches. It is distinguished by its bold white wing stripe. The Sanderling dashes after the retreating waves, walking rapidly like a small windup toy. In summer it is a rust color, but in winter it is pale gray with black shoulders.

Plovers are small, stocky shorebirds that move up and down the beach in short starts and stops. They have thicker necks, shorter bills, and larger eyes than the Sanderlings. The black-bellied plover has a black breast and speckled back during the breeding season in the Arctic, but it is gray in its winter plumage (Figure 7–17). In flight, it has black axillars (wingpits) and a white tail. The ruddy turnstone is a squat, robust, orange-legged shorebird. It gets its name from its unusual habit of turning over stones in search of food. It has a harlequin face and breast pattern during the breeding season, with a pale-orange back. In winter, the pattern is still visible in the dull-brown feathers.

Summary

The beach is inhabited by clams, crabs, beach hoppers, and birds that depend on the waves for their food. The clams dig into the sand and extend their siphon tubes into the overlying waves to filter small organisms out of the water. Crabs raise their antennae to filter out organisms, or scurry about on the sand in search of food. The small beach hoppers and pill bugs live on or within the grains of sand on the beach.

The dunes are held in place by different types of beach grass. The pioneer species become established on the drift line, which marks the farthest advance of storm waves. The tall grasses, American beach grass in the north and sea oats in the south, grow upward as they are buried by sand. Scrub and thickets become established on top of the dunes and replace the beach grasses. Maritime forests and sand flats are formed in the protected areas behind the dunes.

Shorebirds are often abundant along the sandy beaches and dunes. Gulls and terns fly along the shore in search of food in the surf. Sandpipers and plovers run along the beach, snatching food from the waves. Many of the shorebirds migrate along the coast from the Arctic to the tropics, or from the Northern to the Southern Hemisphere and back.

Selected Readings

ART, H. W. *Ecological Studies of the Sunken Forest, Fire Island National Seashore.* New York: Washington National Park Service, 1976. A discussion of the ecology of a maritime forest on a barrier island.

CARSON, R. *The Edge of the Sea.* Boston: Houghton Mifflin Co., 1979. A beautifully written description of the plants and animals that inhabit the coast.

CHAPMAN, V. J. *Coastal Vegetation* (2nd ed.). Oxford: Pergamon Press, 1976. The ecology of coastal vegetation in Britain, northern Europe, and eastern North America.

FOTHERINGHAM, N. *Beachcombers Guide to Gulf Coast Marine Life.* Houston: Gulf Publishing Company, 1980. A profusely illustrated and entertaining guide to the beaches of the Gulf coast.

GOSNER, K. L. *A Field Guide to the Atlantic Seashore.* Boston: Houghton Mifflin Co., 1979. An excellent and well-illustrated field guide to the plants and animals along the Atlantic coast.

LESLIE, C. W. *Nature Drawing: A Tool for Learning.* Englewood Cliffs, N.J.: Prentice-Hall, Inc., 1980. A helpful book on nature drawing by a naturalist and artist.

MINER, R. W. *Field Book for Seashore Life.* New York: G.P. Putnam's Sons, 1950. A guide to the invertebrates along the east coast of North America.

MORRIS, P. A. *A Field Guide to Shells of the Atlantic and Gulf Coasts and the West Indies.* Boston: Houghton Mifflin Co., 1973. A useful guide, with beautiful color photographs, for shell collectors.

MORRIS, P. A. *A Field Guide to Pacific Coast Shells (Including shells of Hawaii and the Gulf of California).* Boston: Houghton Mifflin Co., 1974. A handy guide for west coast shell collectors.

NELSON, B. *Seabirds: Their Biology and Ecology.* New York: A. & W. Publishers, 1979. A well-illustrated book on seabirds from all around the world.

PETERSON, R. T. *A Field Guide to the Birds East of the Rockies.* Boston: Houghton Mifflin Co., 1980. A revised version of the classic guide to the birds.

PETERSON, R. T. *A Field Guide to Western Birds.* Boston: Houghton Mifflin Co., 1968. A guide to the birds west of the Mississippi River.

PETERSON, R. T. *A Field Guide to Birds of Texas and Adjacent States.* Boston: Houghton Mifflin Co., 1968. Includes the coastal birds found along the Gulf of Mexico.

PETERSON, R. T. AND M. McKENNY. *A Field Guide to Wildflowers of Northeastern and North Central North America.* Boston: Houghton Mifflin Co., 1968. A useful guide for dune plants.

RANWELL, D. S. *Ecology of Salt Marshes and Sand Dunes.* London: Chapman and Hall, 1972. An excellent discussion of the ecology of sand dunes, concentrating on the British coast.

RICE, T. *Marine Shells of the Pacific Coast.* Everett, Washington: Ellis Robinson Publishing Co., Inc., 1972. A small, inexpensive guide to Pacific shells, illustrated with beautiful color photographs.

ROBBINS, C. S., B. BRUIN, AND H. ZIM. *Birds of North America.* New York: Golden Press, 1966. A useful guide covering the birds of both the Atlantic and Pacific coasts.

TOWNSEND, C. W. *Sand Dunes and Salt Marshes.* Boston: Dana Estes & Co., 1913. A naturalist's book, well illustrated with beautiful black-and-white photographs.

WOODHOUSE, W. W., JR. *Dune Building and Stabilization with Vegetation.* Fort Belvoir, Va.: U.S. Army Corp. of Engineers, 1978. A discussion of how to build dunes, with a good description of the ecology of different dune grasses.

early morning clamming

EIGHT
Salt marsh and tidal flat ecology

Introduction to Salt Marshes and Tidal Flats

Salt marshes and tidal flats are protected areas, behind barrier islands and spits, that are flooded once or twice each day by the tides. The salt marshes, which are above sea level, support a luxuriant growth of marsh grasses and are only flooded by high tides. The tidal flats, which extend below sea level, are uncovered by low tides and are devoid of large plants, but they do support a good crop of microscopic algae.

In the intertidal area behind the barrier island, salt marshes are covered by cordgrasses, *Spartina,* which hold down the sediment and produce their own substrate, peat, when they die. The salt marshes support a diverse population, which ultimately depends on the cordgrass for their livelihood.

The sand and mud flats between mean sea level and low spring tide level are essentially barren of plant material. Algae grows on the low tide flats, but the flats stand in marked contrast to the productive higher salt marshes and the marine grasses below low spring tide. The tidal flats are the home of oysters and the hunting ground of crabs, but the lack of plants gives them a barren appearance. When the tide is high, several species of fish invade the flats and roam the salt marshes. As the tide moves out, predators from the land and wading birds inhabit the salt marshes and flats. A wide variety of burrowing and crawling invertebrates live on and within the sediment.

The lagoonal or estuarine environments protected by barrier islands are important habitats for marine organisms and are very high in organic productivity. Most of the brackish water areas where there is a significant freshwater input support underwater beds of eelgrass, *Zostera,* and several varieties of floating and attached algae. *Zostera* is a major source of food for the brant. When the eelgrass died out several decades ago, the brant populations crashed until the eelgrass recovered (Figure 8–1). Near the inlets where the salinity more nearly approaches that of sea water, turtle grass, *Thalassia,* increases in importance. The grasses in the lagoonal environment are a source of food for the organisms, but also provide hiding places and substrate for a diverse faunal community. The estuary and salt marshes provide a nursery for many marine organisms that must spend a portion of their life cycle out of the sea to survive. Certain flat fish, rays, and horseshoe crabs are adapted to life on the tidal flats and roam the flats when the tide is in, digging for buried organisms.

Although a stressful environment for many plants and animals, the salt marsh is one of the most productive areas on earth. The high productivity and daily water exchange with the ocean make the lagoons and marshes a vital link with the ocean ecology. Many ocean fish spend their youth in lagoons and tidal flats and enter the sea as adults. Other fish spend their entire life cycle in the lagoons and are preyed upon by ocean fish that migrate up and down the coast. The salt marshes and tidal flats are small in total area, but provide an important resource for fish of the open ocean.

FIGURE 8–1.
Eelgrass, *Zostera*, covers the
bottoms of shallow bays and
collects in rows along the edge of
the marsh. It is the principal food
of the Brant goose and provides
shelter for many small animals.
Drawing by C. W. L.

Salt Marsh Ecology

Salt marshes occupy the high intertidal zone behind the barrier island. Salt marshes contain a thin veneer of organic sediment laid down in the high intertidal area that overlies low intertidal and subtidal muds. The three major salt marsh groups in North America are (1) Bay of Fundy and New England marshes, (2) Atlantic and Gulf coastal plain marshes, and (3) Pacific marshes. About 80 to 90 percent of the Atlantic and Gulf coastal plain contains marshes, but only 10 to 20 percent of the Pacific coast is suitable for marshes.

The Pacific coast is a collision coast, with active mountain building where coastal uplift has limited the amount of space available for marshes (see Chapter One). The coastal plain areas of the Alaskan arctic slope and on Baja, California, are flat enough for the development of extensive marshes. Along the rest of the Pacific coast, marshes are found along the margins of protected bays, such as San Francisco Bay, or in coves along the rocky coast.

Along the Atlantic-Gulf coast, marshes are essentially continuous over broad areas, and originally covered more than 20,000 square kilometers (5 million acres). Georgia alone has close to 1550 square kilometers (360,000 acres), but this is dwarfed by the salt marshes of Louisiana. Although some broad marshes are developed behind spits on Cape Cod, marsh areas on the New England and Canadian coasts occur in protected bays.

FIGURE 8–2. Spartina alterniflora grows along the tidal creeks
in the low marsh at Plum Island, Massachusetts.
Photo by W. T. Fox.

Low marshes   Salt marshes are divided into two fundamental zones, low
marsh and high marsh. On the basis of the flora, low marshes
are younger, occupy a topographically lower position, and are
more marine or estuarine in character. Low marshes generally
develop on mud that contains up to 90 percent interstitial
water, and are flooded at least once each day. The boundary between high and low marshes occurs at about mean high-water
neap tide.

The low marsh, directly above the intertidal flats, is dominated by different ecophenotypes (growth forms) of the smooth
cordgrass, *Spartina alterniflora* (Figure 8–2). Growth forms are
tallest along the edge of the tidal creeks where the grass is the
most luxuriant. Stems of *Spartina* become progressively shorter
toward the high marsh. Taller variants of *Spartina* are more
abundant along the tidal channels of the south than in New
England. However, along brackish streams in the southeast,
Spartina alterniflora is largely replaced by the big cordgrass, *S.
Cynsuroides* (Figure 8–3).

High marshes   High marshes are older, occupy a higher topographic position,
and are more influenced by terrestrial or land conditions than
low marshes. They are submerged only during high spring tides
and may be exposed to the air for up to 10 days between tidal
flooding. High marshes are found on sand that is permeable and
drains, leaving a drier substrate. The marshes of the Mississippi
delta are considered "flotants" because of the nearly liquid
substrate.

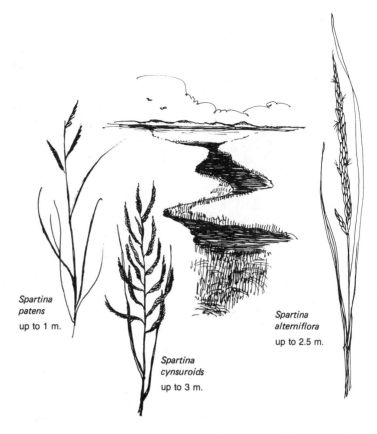

Spartina
patens

up to 1 m.

Spartina
cynsuroids

up to 3 m.

Spartina
alterniflora

up to 2.5 m.

FIGURE 8–3.
Several different species of marsh
grass cover the marshes and grow
along the tide channels.
Drawing by C. W. L.

The high marshes along the Atlantic coast contain a much wider diversity than the low marshes and, with their long intervals of subaerial exposure, more closely resemble terrestrial environments (Figure 8–4). In New England, Barnstable marsh on Cape Cod has been studied by Redfield. Barnstable high marsh is dominated by salt meadow cordgrass, *Spartina patens*, and spike grass, *Distichlis*. Other plants intermingled with the grasses include sea-lavender, *Limonium*; seaside plantago, *Plantago*; aster, *Aster*; seaside goldenrod, *Solidago*; salt brush, *Atriplex*; sea-blite, *Suaeda*; and glass wort, *Salicornia*. Along the margins of the marsh at the higher levels, black-grass, *Juncus*, is common. The salt pannes are often surrounded by samphire, *Salicornia europaea*, and glass wort, *S. bigelowii*. Where fresh water is supplied to the marshes from the groundwater, and where exchange with the sea is reduced by dikes or bridges, the halophytes are replaced by reeds, *Phragmites*; saltmarsh bulrush, *Scripus*; and cattail, *Typha*.

The high marshes along the southeastern coast of the United States are dominated by—in addition to smaller variants of *Spartina patens*—glasswort, samphires, spike grass, and

FIGURE 8–4. The high marsh at Barnstable, Massachusetts, is covered with salt meadow cordgrass, *Spartina patens,* and spike grass, *Distichlis.*
Photo by H. Art.

marsh rush, *Juncus.* In the northeast, *Spartina patens* grows upright, but is laid flat by storm surges that cover washover fans. In the southeast, the *S. patens* remains upright during overwash, and rapidly recolonizes washover fans. Therefore, the overwash process in the southeast is much more frequent because of the lower level of the primary dunes, but the high marsh recovers more quickly in the southeast, because *Spartina patens* grows rapidly through the washover fans. When the overwash sediments move into the bay, the response is the same in the north and south. The low marsh cordgrass, *Spartina alterniflora,* becomes established on the washover fan and creates a salt marsh community.

Insects   Tidal marshes that alternate from wet to dry once or twice each day are breeding grounds for several insects. Green head flies breed in the upper marsh and fly across the dunes and beach. The green heads have the nasty habit of swarming and biting. Black fly traps are placed along the edge of the marsh to help control the green heads. The black traps absorb heat and attract the flies. A funnel-shaped screen allows the green heads to enter the trap but prevents their escape. The traps help to keep the green head population down, but enough green heads avoid the traps to make a mid-summer trip to the barrier beach an unpleasant experience during green head season.

Mosquitoes also breed in the still water left by high tides on the marshes. The mosquitoes are controlled by draining the higher marshes. A rectangular pattern of drainage ditches indicates a mosquito-control project. When a high spring tide floods the upper marshes, the channels rapidly drain off the excess water and cut down on the mosquito breeding areas.

Snakes

Snakes also live in the marshes and slither up on the barrier islands. The most dangerous snake encountered along the marsh is the cottonmouth, *Agkistrodon*. The cottonmouth is a water moccasin, related to the other pit vipers, rattlesnakes, and copperheads. Although cottonmouths are present on the Cape Hatteras National Seashore, they seem to avoid people and will disappear unless they are pestered. The cottonmouth may grow 2 meters long, but they are usually less than a meter. They are brown with indistinct black bands, and have a yellow belly with dark bands. They also have a dark band that runs from the corner of the mouth to the eye. It is distinguished from nonpoisonous water snakes by its deep spade-shaped head, white mouth, and light lips. It lives on the bay side of the marshes on southern barrier islands and eats fish, frogs, and other amphibians.

Most snakes encountered in the marsh are harmless, but the reputation of the cottonmouth has caused many to be killed. If a snake has a spade-shaped head, keep your distance to avoid the venom. However, if it does not have a spade-shaped head, let it live in peace in the marsh where it helps keep the balance of nature.

Tidal Flats and Lagoons

On most low coasts, unvegetated tidal flats are found between low spring tide and the fringe of the salt marsh. Although tidal flats are generally barren of higher plant life, microalgae are extremely abundant. The lagoons are constantly submerged below low spring tide. The flats and lagoons are the home of a wide variety of different organisms that live in close association with the bottom. The benthic animals include the epifauna that live fixed on top of the substrate, the infauna that live buried in the sediment, and the mobile epibenthos that move about on and in the sediments, foraging and scavenging for food. The marine epifauna and infauna are almost solely invertebrates, while the epibenthos include both invertebrates and vertebrates.

Much of the following material on the ecology of the flats was gleaned from the excellent U. S. Fish and Wildlife booklet, *The Ecology of Intertidal Flats of North Carolina: a Community Profile*, by Charles and Nancy Peterson. The booklet is a comprehensive summary of current information on the tidal flats of southeastern United States.

Microscopic plants

Although an intertidal flat does not contain marsh plants or sea grasses, macrophytic algae are frequently found attached to shell debris, pebbles, and small fragments of hard substrate. These small plants are best studied under the microscope, but they are obvious to the naked eye by the discoloration on the flats. Benthic diatoms appear as brownish stains on the mud and occasionally form dense, multilayered sheets on the intertidal flats. Mats of blue-green algae occasionally tint the sediments dark green in the high intertidal zone. During low tide, the blue-green algal mats often dry into a hard black crust that resembles asphalt. Even where they are not conspicuous, benthic microalgae are important primary producers on intertidal flats.

Along the southeast Atlantic coast, filamentous brown algae are abundant on high-salinity sand and mud flats. In spring, massive amounts of filamentous green algae replace the brown algae. In North Carolina, the greens are abundant from February through June, and from April through July; the leafy green alga *Ulva* covers up to half the total area of the flats. Bacteria and fungi are abundant on the surface of sediment particles on the flats. The smaller particles have a greater surface area and support a larger population of microbes than the coarser sediments. Both bacteria and fungi convert dead organic matter into inorganic nutrients for the primary producers (plants) and consumers (animals) in the lagoon. Fungi aid in decomposing plant material, and bacteria are abundant decomposers of both plants and animals. Fungi work inside the detrital particle by extending long hyphae, while bacteria generally act on the surface of the particle. The process of breaking down dead organic matter into inorganic nutrients is called mineralization.

In addition to their role as mineralizers, bacteria and fungi also serve as a trophic link between the relatively indigestible parts of the plants, including cellulose and lignin, and the potential plant consumers. Most of the marsh plants and some of the sea grasses are not grazed significantly by herbivores. More than half of the dead plant material from marsh plants and sea grass is carried away and ultimately processed by decomposers elsewhere in the estuarine and coastal systems.

At high tide, when the intertidal flats are covered by flood waters, phytoplankton grow and reproduce in intertidal areas. Phytoplankton populations, dominated by various diatoms, especially *Skeletonema,* are abundant in the summer months. The area of the euphotic zone where phytoplankton thrive is greatly expanded at high tide. During winter, the water on the tidal flats is clear and the phytoplankton population is low. In North Carolina estuaries, phytoplankton account for almost half of the total productivity, while benthic microalgae contribute less than one-tenth, and Zostera only about 1 percent of the productivity.

Tidal flat animals   Four different types of feeding mechanisms are represented on the tidal flats: (1) suspension feeders, (2) deposit feeders, (3) predators, and (4) scavengers. The suspension feeders filter microscopic food particles out of the water column. Clams pump water in through the body cavity and out through the gills, where food particles are collected. Their diet consists of phytoplankton, resuspended benthic algae, and detrital particles with microbiotic organisms on their surface. The deposit feeders ingest bottom sediments and assimilate living microalgae, as well as bacteria and fungi, on the detrital particles. Although some deposit feeders are selective as to particle size, most just plow through the sediment. Some deposit feeders construct tunnels as they turn over the sediment. The process of reworking the sediment and eating the microbiota is known as bioturbation. The predators ingest living animals when the prey is relatively large. The predators are considered macrophages if the prey is large relative to the body size of the predator, and microphages if the prey is small relative to the predator. While most of the microphages would be considered deposit feeders, the macrophages are the more common predators, including various types of worms, polychaetes, bloodworms, and ribbon worms. Scavengers roam the bottom and eat the dead animals before they decay.

Suspension feeders. Suspension feeders on the tidal flats filter microorganisms from the tidal currents. The suspension feeders include several types of worms, mussels, oysters, clams, and shrimp.

The most abundant suspension feeder on the sand flats in the southeast is the worm, *Balanoglossus*, which is ¾ meter long. On the flats of North Carolina, the worm has a density of 4 per square meter and processes about 140 cubic centimeters per day during the warm months. *Balanoglossus* lives in a cone-shaped feeding funnel that is surrounded by coiled extrusions of feces.

The lug worm, *Arenicola*, which is about half a meter in length, lives fairly deep in the sand, with the burrows extending below 30 centimeters. The tubes occupied by *Arenicola* can be identified by the gelatinous egg cases that trail off in the current.

The blue mussel, *Mytilus*, is the most apparent suspension feeder on the tidal flats in Europe and the Pacific coast of North America. However, the Atlantic oyster, *Crassotrea virginica*, is the most obvious suspension feeder in the southeast Atlantic states and along the Gulf coast. The oysters are found in dense clusters at the high intertidal zone at the lower edge of the salt marsh (Figure 8–5). More oysters are also found on the lower level of the shoreline, where they tend to be subtidal. The oysters feed on suspended algae and phytoplankton with some resuspended benthic diatoms. Along the mid-Atlantic states, seed oysters are planted in mounds and harvested along the flats.

Oysters have been successfully transplanted from the Atlantic to the Pacific, and from Japan to North America. Following the gold rush of 1849, Atlantic oysters were packed in barrels with the intertidal mud and carried to San Francisco by train. When they were transplanted in San Francisco Bay, they

FIGURE 8–5.
The Atlantic oyster, *Crassotrea
virginica*, grows in clusters.
Drawing by C. W. L.

lived but did not reproduce rapidly. Next, the Japanese oyster, *Crassotrea gigas,* was introduced into North America from British Columbia, through Washington and Oregon, to California. The Japanese oyster thrived on the west coast of North America, but many other species that were brought along with the oysters also grew and took hold along the coast. The Columbia oyster, *Ostrea lurida,* the only native oyster on the Pacific coast of North America, represents only a small fraction of the total marketed oyster crop. The well-known Willapa Bay oyster from the Washington coast is a transplanted Japanese oyster.

Several varieties of clams are collected from the sand bars or mud flats near the mouths of the tidal inlets. The quahog is one of the most sought-after clams to accompany a lobster dinner. The northern quahog, *Mercenaria mercenaria,* has an oval shape with concentric growth lines and has a length of about 10 centimeters (4 inches) (Figure 8–6). It is also known as the little-neck clam, cherrystone, or round clam. Its southern relative, *Mercenaria campechiensis,* is found south from Chesapeake Bay along the Atlantic coast. The southern quahog is about 15 centimeters (6 inches) long and has a purple stain along the inner margin of the shell. The quahog has a thick shell that provides protection from all but the most persistent predators, except man.

In contrast to the quahog, the soft-shell clam, *Mya arenaria,* has a very fragile shell. The soft-shell clam is also known as the steamer clam because it is used to fill pots for a clam feast. The soft-shell clam has the distinctive habit of shooting up a small jet of water when disturbed, which has earned it the vulgar name, pissclam.

The intertidal mud flats contain large numbers of short-lived opportunistic species of Polychaete worms and amphipods. The population density of infauna on the mud flats reaches a peak in the spring, and declines in the summer when predation is high and reproductive activity is low. The small species of razor clam, *Tegelus,* is less abundant, but makes up 90 percent of the biomass on the mud flats in North Carolina. On the muddy

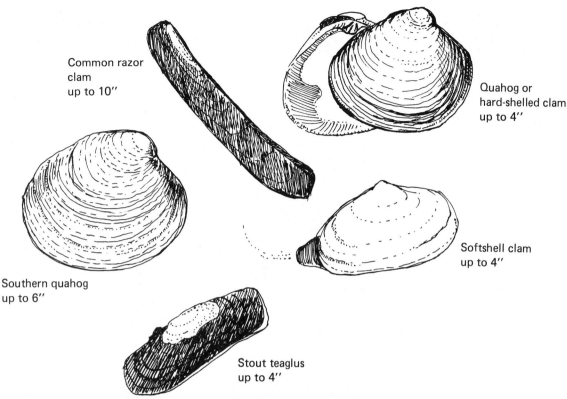

Common razor
clam
up to 10"

Quahog or
hard-shelled clam
up to 4"

Southern quahog
up to 6"

Softshell clam
up to 4"

Stout teaglus
up to 4"

FIGURE 8–6.
A group of clams found in the tidal flats: common razor
clam, *Ensis*; Quahog, or hard-shell clam, *Mercenaria mercenaria*; soft-shell clam, *Mya*; stout razor clam, *Tegelus*; and
southern Quahog, *Mercenaria campechiensis*.
Drawing by C. W. L.

sand flats, the mud shrimp, *Upogebia,* is the dominant infaunal
organism. Both *Tegelus* and *Upogebia* are suspension feeders
that live in semipermanent burrows.

On the sand flats near the mouths of tidal inlets where
strong tidal currents sweep by, sea pansies, *Renilla,* are abundant. Sea pansies are colonial coelentrates that form a purple
disk about the size of a silver dollar. The sea pansies are anchored in the sand by a pedicle so they could also be considered
infauna suspension feeders. They are also abundant on the open
beach in the surf zone.

Deposit feeders. Deposit feeders live on the mud flats
where they ingest mud and eat algae, bacteria, and fungi on the
bottom sediments. The major deposit feeders include different
species of crabs, shrimp, sand dollars, sea urchins, and worms.

Fiddler crabs, *Uca,* are deposit feeders that scurry around
on the tide flats, looking comical with their oversized claws. It

is thought that the large claw is used by the male to beckon a prospective female mate (Figure 8–7). The fiddler crabs construct burrows with round entry holes near the high-tidal marsh. The holes are surrounded by piles of spherical droppings. They forage the intertidal zone at low tide, ingesting benthic algae and detritus. The stone crab, *Menippe,* which has a distinctive claw, and the spider crab, *Libinia,* also forage in the low intertidal zone.

The horseshoe crabs, *Limulus,* dig pits for clam larvae and other infaunal organisms from April through October, then abandon the tidal flats for deeper water. The crabs resemble horse hooves with sharp tails (Figure 8–8). The horseshoe crabs are not true crabs, but are more closely related to the scorpions, spiders, and ticks.

The horsehoe crabs have survived almost unchanged since the Devonian Period about 360 million years ago and are considered living fossils. There were once several different species of horseshoe crabs, but only four species carry on the line, three in southeast Asia and one along the Atlantic coast from Maine to Yucatan. The crabs are most abundant along the shore of Delaware Bay. Like the grunion in California, the mating and egg laying of the horseshoe crabs are timed to coincide with the high spring tides. Hundreds of thousands of horseshoe crabs come on shore during the spring to spawn, and then return to the bay.

Several varieties of shrimp are deposit feeders in the marine grasses just below the tidal flats. The grass shrimp, *Palaemonetes,* and the brown shrimp, *Paneus,* live in vegetated areas, but start their life in shallow, muddy tidal creeks. At the end of summer, they migrate to the sand flats and eventually out to the open sea. In the estuary, they live in microbial flora, or *Spartina* detritus. The mysid shrimp, *Neomysis,* is found on the flats south of Chesapeake Bay.

Two deposit feeders, the sand dollar, *Mellita,* and the heart urchin, *Moira,* plow through the bottom sediment. The

FIGURE 8–7.
Male fiddler crab, *Uca,* signaling to prospective mate near his burrow hole (length: 5 cm).
Drawing by C. W. L.

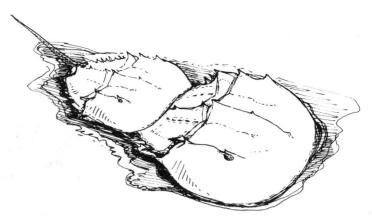

FIGURE 8–8.
Female horseshoe crab, *Limulus,*
with smaller male riding her
back. Horseshoe crabs mate on
the beach at high spring tide.
Drawing by C. W. L.

tube-building polychaetes, *Amphitrite* on the mud flats, and
Diopatra on the sand flats, are surface deposit feeders. Gastropods including the tulip snails, *Fasciolaria,* also feed on the
benthic infauna.

Suspension feeders are usually not abundant in the same
area as dense deposit feeders. The deposit feeders stir up the
bottom and create a loose mud layer, which clogs the gills of
suspension feeders. The deposit feeders also limit the influx of
new species by eating larvae shortly after they have settled. Suspension feeders dominate on sand flats, and deposit feeders
dominate on mud flats.

Predators. The mobile epibenthos, those animals that are
the predators on the flats, include crabs, whelks, and shrimp. In
the mid-Atlantic states, the blue crabs, *Callinectes,* are major
predators on the benthic infauna at high tide. The thin-shelled
clams, *Macoma* and *Mya,* and the juvenile hard clams, *Mercenaria,* are included in the diet of the blue crabs. The blue crabs
with a tolerance for low salinity move far up the estuaries into
brackish water to seek their prey. The adult crabs, which prefer
the deeper water of the tidal channels, also forage on the tidal
flats in the spring. In the summer, blue crabs become inactive
and retire to deeper water during the day, while the young crabs
forage on the exposed flats. During winter, from December to
March, the blue crabs of North Carolina are rarely on the flats;
the crabs remain in the deep channels to escape the extreme
cold. The blue crabs leave small pits, about 8 to 10 centimeters,
on the flats where they have been foraging in the mud.

Three species of whelks, which are large gastropods or
snails, are predators on the lower margin of the mud and sand
flats of North Carolina. The channeled whelk, *Busycon canali-
culatum,* lives on carrion and is often found in crab pots. The
knobbed whelk, *Busycon carica,* and the lightening whelk,
B. contrarium, are voracious eaters of the hard clam, *Mercenaria,* and the dog clam, *Chione* (Figure 8–9). The whelks pry
open the clam shell by rasping away until a gap is opened in the

FIGURE 8–9.
The knobbed whelk, *Busycon
carica* (length: 22 cm).
Drawing by C. W. L.

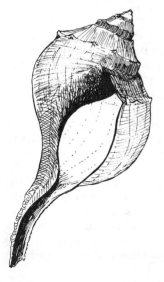

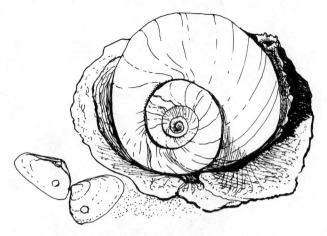

FIGURE 8–10. The moon snail, *Polinices*, drills a hole through a clam shell with its radula (length: 10 cm). *Drawing by C. W. L.*

margin. The whelk leaves the clam shells articulated when the meal is complete.

The moon snail, *Polinices*, also has a varied diet that includes clams. Small round holes that are found on empty clam shells are largely the product of the moon snail, which uses its radula, a coiled strip of teeth, to drill a hole through the live clam (Figure 8–10). Many of the moon snail shells provide homes for the omnivorous hermit crabs, *Pagurus* and *Petrochirus*.

The snapping shrimp, *Alpheus*, is a predatory shrimp on the mud flats. The mantis shrimp, *Squilla*, are voracious predators that live in permanent burrows that extend fairly deep in the sediment. They emerge from the burrow when a prey passes by, and grab it with their powerful claws, or shelae.

Scavengers. The scavengers are relatively mobile species that move to any dying or dead animal matter. The gastropods of the Nassariidae family, including the mud snail, *Ilyanassa*, are generally scavengers, but they take in a large amount of sediment, so they could also be considered deposit feeders.

*Fish on
the tidal flats*

The fish and shorebirds play complementary roles on the tidal flats. When the tide is in, the fish are the major predators roaming over the flats and digging in the soft bottom. When the tide is out, the shorebirds move on the flat and take over from the fish. Four different habits are displayed by the fish on the tidal flats: (1) planktivores, (2) detritivores, (3) predators on the benthic infauna, epifauna, and small epibenthos, and (4) predators on fish and larger mobile epibenthos. Many of the fish live and feed on the tidal flats during high tide at some time during their life cycle, and other fish are dependent on prey that have lived on the flats.

Plankton feeders. The planktivores that take up residence on the tidal flats include the anchovies, menhaden, herring, silversides, and killifish. The striped anchovy, *Anchoa hepsetus,* are common on the flats in the summer and spend the winter months at sea. The Bay anchovy, *Anchoa mitchilli,* are abundant on the flats year round. The anchovy eat mainly small shrimp, small bottom-dwelling mollusca, and zooplankton while they are on the flats. The Atlantic menhaden are the most abundant fish along the east coast. In the summer, they live on the tidal flats from Maine to Florida. Their diet consists of mainly phytoplankton, which are abundant on the shallow flats, along with some zooplankton and resuspended detritus. The adult menhaden are about a foot long with a large head. Although the menhaden provide the largest catch to the United States, they have an oily flesh and do not taste good. The menhaden are used as fertilizer and for fodder, and the oil is mixed with paint and varnish as a drying agent (Figure 8–11).

The silversides and killifish are year-round residents of the tidal flats and eat small crustaceans, polychaetes, and detritus (Figure 8–11). The striped killifish, *Fundulus majalis,* is the most important killifish in the unvegetated flats, which are sand and have a higher salinity. On the muddy flats among the vegetation and in the deeper pools, the sheepshead minnow, *Cyprinodon,* and the mummichog, *Fundulus heteroclitus,* are more abundant. The Atlantic silversides, *Menidia,* which eat zooplankton, are abundant on the flats year round. The tidewater silversides, *Menidia beryllina,* and the rough silversides, *Membras martinica,* are common year round, but they are found on the brackish water flats farther from the tidal inlets.

FIGURE 8–11. The silversides, menhaden, flounder, and southern stingray are common on the tidal flats.
Drawing by C. W. L.

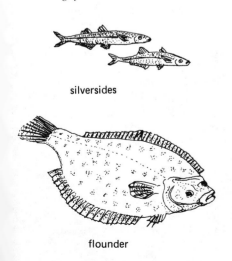

silversides

flounder

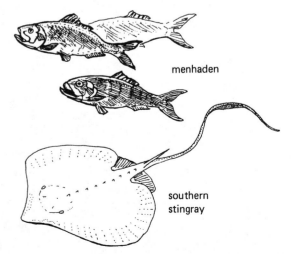

menhaden

southern
stingray

Detritus feeders. The mullets, *Mugil,* are almost pure detritus feeders. They live on organic detritus, epiphytic algae, and littoral diatoms. The mullets grow up in the estuaries and lagoons, and migrate south in massive schools during the late fall. In the mid-Atlantic states, the first real cold front in the fall is called a "mullet blow" and coincides with the start of the mullet migration. The mullets migrate close to shore, and their dark forms can be seen in the breaking waves along the surf zone. In Florida, mullets are caught in schools by casting nets from the bridges over tidal inlets.

The pinfish, *Lagodon,* and the file fish, *Monacanthus,* are abundant on the tidal flats and eat mainly encrusting green algae, sea grasses, small fish, mollusks, and shrimp. The file fish lives on detritus and small crustaceans, including copepods and hermit crabs.

Predators. Several different types of predators swim across the tidal flats at high tide. The rays and skates are present from spring through the fall, when they dig pits in the sediments in search of food. The flapping wings are the fleshy part of the pectoral fins, which enable the rays to excavate craters. The cownose ray, *Rhinoptera,* eats mostly commercial shellfish, including clams, scallops, and oysters, so it is not appreciated by the local shell fishermen. The most abundant ray, the Atlantic stingray, *Dasyatis,* does not excavate pits and lives on crustaceans and fish (Figure 8–11). In Europe, ray fins are cut into pieces that closely resemble scallops in taste and appearance.

The flatfish, including flounder, sole, and tonguefish, forage on the bottom for small fish and shrimp. The flounders lie on the bottom with their right side down and the right eye moved alongside the left eye. The pattern of color blotches and marks on the upper surface of the flounder blends in with the surrounding bottom, providing a perfect disguise. The summer flounder, *Paralichthys dentatus,* and the southern flounder, *P. lethostigma,* prey on small silversides and anchovy along with shrimp and crab. The hogchoker, *Trinectes,* is a sole that eats annelid worms and small crustaceans along with detritus. Flounder are year-round residents of the lagoons until they reach about 18 months; then they leave for the open ocean and return to spawn. Flounder are often caught by gigging, or using a barbed spear to gig the fish on the shallow bottom while walking or poling in a small boat. Many fish that are normally considered marine fish are bottom-feeding fish during their juvenile stages. These include spot, *Leiostomus;* Atlantic croaker, *Micropogonias;* silver perch, *Bairdiella;* and pigfish, *Orthopristis.*

Although they are common in the warm months, most of the adults migrate offshore to spawn in the fall and winter. Most of their feeding takes place outside of the tidal flats. The lizardfish, *Synodus,* and the oyster toad, *Opsanus,* are sometimes abundant on the flats, but spend most of their time at sea.

Sharks, particularly the Atlantic sharpnose shark, *Rhizopriono-don,* are often seen in the shallow waters eating fish and crabs. The larger sharks, which may enter the tidal flats at night, are the end of the fish food chain on the flats.

Seasonal patterns in fish. There are large seasonal variations in fish populations on the tidal flats. In early spring, the North Carolina flats are populated by silversides and killifish. In March, April, and May, large schools of Atlantic croaker, spot, menhaden, and flounder appear, along with juvenile striped mullet. In late spring large schools of young silversides enter the lagoons. Small pinfish inhabit the grassy areas in the spring. The diversity of fish is highest in the spring and fall when fish are migrating through the area. During the summer, the baitfish, flatfish, and rays are most abundant. In early fall, the bluefish come in large schools from the north; striped mullet form schools and head south; and spotted sea trout and kingfish, *Menticirrhus,* enter the shallow water. The flounders leave in late fall, with the killifish and silversides left behind to tend the lagoon during the winter.

Marsh and tidal flat birds

A walk along the edge of a salt marsh with a pair of binoculars provides an opportunity to observe several different types of birds. Wading birds with stiltlike legs and slender necks hunt for fish near the tidal inlets. Deep-probing shorebirds extend their curved beaks into the mud in search of worms and clams. Diving and floating birds swim underwater to gather plants and animals along the bottom. Large birds of prey, such as osprey and marsh hawks, which nest in marsh islands, circle overhead and dive for fish and other prey.

Wading birds. The medium to large wading birds, including the herons and egrets, have long, spindly legs, long necks, and spearlike bills. They stand with their heads erect, ready to dart into the water after small fish. The Great Blue Heron is a lean, gray bird that stands 120 centimeters (4 feet) tall. In flight, its neck is folded into an *S* and its legs trail behind. The Louisiana Heron is a smaller, dark heron with a contrasting white belly and a white rump. The Great Egret is a large, slender, white heron with a yellow bill and black legs and feet. The Snowy Egret is a smaller, white heron with distinctive golden slippers.

The herons are typically found in marshes on shallow tidal flats, but they will also fish in shallow ponds along a beach. Several of the herons migrate north and south across the equator, wintering in South America and breeding in the marshes, swamps, and tidal flats of eastern North America. The herons are colonial nesters with their rookeries on natural and spoil islands in the lagoons behind barrier islands. Where tidal inlets have been dredged, the dredge-spoil islands are higher than the natural islands and are less frequently subjected to flooding and overwash. For example, in North Carolina, more than 60 percent of the estuarine islands are artificial, composed of dredge spoils. Egg predators, such as raccoons, foxes, and rats, are usually absent from the spoil islands. Most of the spoil islands are located near inlets where abundant fish are available for the wading birds.

Deep-probing shorebirds. The deep-probing shorebirds reach deeper into the sediment with their bills in search of large invertebrates. The Long-billed Dowitcher is a snipelike bird of the open mud flats with a very long bill (Figure 7–17). The sides of the breast are barred, and it has a reddish belly. The Whimbrel is a larger gray-brown sandpiper with a long, downward-curving bill.

The Willet is a large, gray sandpiper with a distinctive black-and-white wing pattern in flight. Although the Willets generally inhabit the marshes and mud flats, they commonly feed on the beaches, where they go after the mole crab, *Emerita,* near the high-tide line.

The Greater Yellowlegs is a slim, gray sandpiper with bright yellow legs. In contrast to the Whimbrel, it has a long, upturned bill. In flight, it is dark winged and, unlike the Willet, has no wing stripe. It is a small wading bird, common in the open marshes and mud flats.

The Marbled Godwit is one of the larger sandpipers with a very long and slightly upturned bill. It is common in the marshes, where it is noted by its mottled, buff-brown color with cinnamon wing linings.

Most of the deep-probing shorebirds are abundant on both the east and west coasts of North America. For example, the Whimbrel and Long-billed Dowitcher are present on both coasts, but they are more abundant on the Pacific, where they feed on the ghost shrimp, *Callianassa,* and the mud shrimp, *Upozebia,* which are quite abundant.

Diving and floating birds. The diving and floating birds forage while floating. The fish-eating birds include loons, grebes, mergansers, and comorants, which prey on large fish. The diving birds, including scaup, eider, scoters, bufflehead ducks, goldeneye, ruddy duck, redhead, and black duck, dive to the bottom after mollusks and other benthic fauna. The geese, brant, swans, canvasbacks, and most other ducks eat some invertebrates, but subsist mainly on bottom vegetation. Although most of these birds head to the Arctic for breeding in the summer, their wintering grounds are the coastal areas of the southeast, from Chesapeake Bay to the south.

The common loon is a large low-swimming bird with a stout daggerlike bill (Figure 7–17). In the summer it has a checkered back, while in winter, it is dark above and whitish below. In the summer it nests on inland freshwater ponds and in winter it seeks the open salt water. This is true also of the marsh and bay ducks that can only be seen along the coast from fall until early spring.

The Canada goose is the most widespread goose in North America (Figure 8–12). Canada geese travel in flocks with a long *V*-formation. The deep musical honking or barking can be heard far overhead as the geese migrate south in the fall. When they are on the ground feeding, at least one goose always has its head raised, watching for signs of an intruder.

The bufflehead duck is a small duck that is often seen along the coast in winter. The male has a white body with a black back (Figure 7–17). Its puffy, black head is adorned with a large, white, bonnetlike patch. The female is dark and compact, lacks the bonnet, but has a white cheek spot.

The common eider duck is a northern sea duck that comes to nest in summer on rocky coasts and shoals only as far south as Maine, and in winter is seen in bays down to southern Massachusetts. The male eider has a black belly and a white back with a black crown (Figure 7–17). The male goldeneye duck is a white duck with a black back and a puffy, green head with a

FIGURE 8–12. "Dark Sky—Canada Geese" by Maynard Reece.
Courtesy of Mill Pond Press, Inc., Venice, Florida.

white spot. It nests in tree holes in inland lakes, and winters along the entire North American coastline. In early spring, the courtship postures and pranks of the bufflehead and goldeneye are humorous sights to watch, as they bob and splatter about, chasing potential mates. Once paired in April, they will head north to begin nesting.

Birds of prey. The large birds of prey reign over the tidal flats. The magnificent osprey soars at 60 meters above the water, then plummets to capture large fish, with its talons (Figure 7–17). The osprey, also known as the fish hawk, has a spectacular method of catching fish. When it sights its prey, the bird hovers momentarily, heavily beating its wings, then plunges downward like a feathered spear. The fish is seized with the talons, and the bird hits the water with great force, almost submerging. When the osprey rises, it shakes the water from its plumage like a retriever coming ashore, and shifts the prey so that it heads into the wind. The osprey starts its dive from 10 to 30 meters (30 to 100 feet) above the water, and dives with a roar as the wind rushes through its wings. The osprey is encouraged in many fishing areas where nesting sites are provided. The fishermen consider that the osprey help in locating schools of fish that would otherwise be missed.

The bald eagle lives along the coast on an almost exclusive diet of fish. It is somewhat larger than the osprey, and has been known to relieve the osprey of its catch. However, generally, the bald eagle catches its own fish. Both osprey and bald eagle numbers have been reduced due to intake of fish with high pesticide levels, Marsh hawks and short-eared owls hunt for small animals on the high marshes and barrier islands.

Summary

The ecology of salt marshes and tidal flats is closely intertwined with the tides. The marshes that exist above the mid-tide level are flooded only during high tides. The tidal flats that are below mid-tide are exposed by low tides.

Salt marshes are divided into low and high marshes on the basis of the flora. The low marsh, which is developed on mud, is dominated by the tall, smooth cord grass, *Spartina alterniflora.* The high marshes found on sand are covered by short salt meadow cord grass, *Spartina patens,* or spike grass, *Distichlis spicata.* Low marshes are flooded at least once each day, while high marshes are dry except during the high spring tides.

Tide flats cover the area between low spring tide and the edge of the salt marsh. Tidal flats are barren of higher plants but contain abundant microalgae. Animals on the flats are suspension feeders, deposit feeders, predators, and scavengers. The suspension feeders, which filter microorganisms from the water, include worms, mussels, oysters, clams, and shrimp. The deposit feeders, which extract algae, bacteria, and fungi from the bottom sediments, include crabs, shrimp, sand dollars, sea urchins, and worms. The predators on the flats are crabs, whelks, and shrimp. Snails are the major scavengers, eating dying or dead animal matter.

When the tide is in, fish invade the marshes and flats, eating vegetation, plankton, and tidal flat animals. Wading birds hunt for fish near the tidal inlets, and probing shorebirds search for animals in the mud. The diving and floating birds forage for fish, mollusks, or bottom vegetation in the lagoons and marshes.

Selected Readings

GIBBONS, E. *Stalking the Blue-eyed Scallop.* New York: David McKay Co., 1964. A delightful guide to edible seashore life.

GREEN, J. *The Biology of Estuarine Animals.* Seattle: University of Washington Press, 1968. An excellent discussion of estuarine plants and animals around the world.

GROSNER, K. L. *Guide to Identification of Marine and Estuarine Invertebrates.* New York: John Wiley and Sons, 1971. A specialist's guide to the inhabitants of the marsh and tidal flats.

GROSNER, K. L. *A Field Guide to the Atlantic Seashore.* Boston: Houghton Mifflin Co., 1979. A useful companion for crab identification on the tidal flats.

KENNEDY, V. S. (ed.). *Estuarine Perspectives.* New York: Academic Press, Harcourt Brace, 1980. A collection of papers on ecology of estuaries and tidal flats.

HINDS, H. R. AND. W. A. HATHAWAY. *Wildflowers of Cape Cod.* Chatham, Mass.: Chatham Press, 1968. A helpful guide for identification of sand dune and salt marsh plants.

LIPPSON, J. *The Chesapeake Bay in Maryland, An Atlas of Natural Resources.* Baltimore: The Johns Hopkins University Press, 1973. An excellent atlas of the Chesapeake Bay, with a map showing distribution of temperature, salinity, and invertebrates in the Bay.

PERKINS, E. J. *The Biology of Estuarine and Coastal Waters.* London: Academic Press, 1974. A text on the ecology and biology of marshes and tidal flats.

PETERSON, C. H. AND N. M. PETERSON. *The Ecology of Intertidal Flats of North Carolina: A Community Profile.* Slidell, La.: U.S. Fish and Wildlife Service, Office of Biological Services, 1979. An excellent introduction for the general public to the biology and ecology of tidal flats.

RANSWELL, D. S. *Ecology of Salt Marshes and Sand Dunes.* London: Chapman and Hall, 1972. Several interesting chapters on salt marsh ecology.

STERLING, D. *Our Cape Cod Marshes.* Orleans, Mass.: Association for the Preservation of Cape Cod, 1976. A good description of the ecology of the Cape Cod marshes.

TEAL, J. AND M. TEAL. *Life and Death of the Salt Marsh.* New York: Ballantine Books, 1969. A historical account of salt marsh development on the Atlantic coast.

WILEY, M. (ed.). *Estuarine Processes, Volume I, Uses, Stresses and Adaptation to the Estuary.* New York: Academic Press, 1976. A collection of ecologic papers on salt marshes, tidal flats, and estuaries.

WILEY, M. (ed.). *Estuarine Processes, Volume II, Circulation, Sediments and Transfer of Material in the Estuary.* New York: Academic Press, 1976. A collection of papers on the physical and chemical processes in estuaries.

Big Sur Coast. California

NINE
The rocky coast

Introduction to The Rocky Coast

The rocky coast has some of the most beautiful and spectacular scenery in the world. The surf smashes against cliffs, producing great fountains of water that spray the upper reaches of the rocks. Resistant layers of rock form ledges and pools that trap the salt water spray. Waves slowly erode cliffs, leaving unusual forms and shapes in their wake.

One of the striking features of the rocky coast is the zonation of plants and animals. At low tide, organisms can be seen covering the rocks in parallel bands. A black zone of algae and lichen is present above high tide, where small periwinkles graze on the plants. In the mid-tidal zone, barnacles and mussels compete for space on the exposed rocks. Below the low-tide level, many large algae, including laminaria or kelp, sway back and forth in the waves. A wide variety of small organisms live among the larger plants and animals on the rocks and in the tidal pools.

Several different groups of organisms have evolved mechanisms for holding onto the rocks and have adapted to the stressful life in the intertidal zone. The mollusks have developed a powerful foot to hold onto the rocks' surfaces. The barnacles and limpets have evolved protective coverings to shield them from the force of the waves. The starfish and sea urchins have tube feet that act as small suction cups for grasping the smooth rock surfaces.

Geology of The Rocky Coast

The character of rocky coasts varies considerably, from the low, rolling headlands and bays on the coast of Maine to the cliffs along the Pacific coast. In Maine, wooded headlands extend seaward and are separated by long bays and hidden coves. Small islands and picturesque harbors are found all along the coast. In Washington, Oregon, and California, the coast is relatively straight with high cliffs and broad beaches exposed at low tide.

The difference in character of the Atlantic and Pacific rocky coasts can be accounted for by the rock type, tidal range, wave exposure, and longshore currents. Igneous rocks, including granite and basalt, are exposed along the Maine coast. Along the Pacific coast, sedimentary and metamorphic rocks form the coastline. Both Maine and the Pacific northwest have a fairly high tidal range, about 3 meters, but Maine has a semidiurnal tide, while the tide in the Pacific is mixed, with one high-high and one low-low tide each day. The waves on the Pacific coast are much higher during the winter, up to 8 meters, while the waves on the Maine coast reach 2 to 4 meters. In the Pacific, longshore currents sweep sand parallel to the coast and form wide beaches. On the Maine coast, waves are refracted by the protruding headlands, and sand is swept into the bays, forming spits and bars.

Igneous and metamorphic coasts

The coast of Maine is characterized by gently rolling granite headlands that are separated by long, narrow bays. Thousands of small islands and several large islands are scattered along the coast. The geologic features typical of the Maine coast are well displayed on Mount Desert Island in Acadia National Park (Figure 9–1).

Mount Desert Island consists of 3 long, low mountains: Cadillac Mountain (466 meters high), Penobscot Mountain (364 meters) and Benard Mountain (326 meters). The mountains were formed by the coarse-grained, igneous rock, granite, which intruded into a metamorphic rock schist.

About 450 million years ago, New England was covered by a shallow sea that deposited sand, silt, and mud, eroded from a nearby continent. The sea was part of an earlier Atlantic Ocean formed by the splitting of a large continent called Pangaea (see Chapter One). When the continental plates started to move together, about 400 million years ago, the early Atlantic Ocean closed and the continents collided, forming a chain of mountains. The ocean sediments were folded, heated, and compressed into metamorphic rocks. In the roots of the mountain range, granite was intruded into the metamorphic rocks. For the past 325 million years, the region was slowly uplifted and

FIGURE 9–1. Rocky coast of Maine—
Bass Harbor Head, Acadia National Park.
Drawing by C. W. L.

eroded. Eventually the granite reached the surface of the earth and formed the mountains on Mount Desert Island.

During the Pleistocene glaciation, which ended about 10,000 years ago, Mount Desert Island was covered by about 2 kilometers of ice. The glacial ice, which moved southward over the low mountains, gouged, scraped, and plucked granite from the mountain tops. In the low passes between the mountains, the glaciers eroded deep *U*-shaped valleys. When the glacier melted, ponds and lakes were formed in the long valleys. With the rise in sea level, one valley, Somes Sound, was flooded to form a fjord. The hills left by the glacier became rounded headlands, and small valleys formed coves or bays (Figure 9–1).

With the retreat of the glacier, the processes of weathering and wave action took over along the coast. Granite is formed deep within the crust at high temperatures and under tremendous pressure. When the overlying rocks are eroded and the granite is slowly uplifted to the surface, the rock cools and the pressure is relieved. When the pressure is removed, the granite expands and cracks, in a process called exfoliation. Curved sheets of granite are spalled off the surface, and deep cracks called joints extend into the granite.

FIGURE 9–2. A weathered basalt dike forms a slot in the rock at Taylor's Head, Nova Scotia.
Photo by W. T. Fox.

In some areas, fine-grained basalt, a dark, igneous rock, moves as a fluid through the cracks. The basalt hardens to form dikes that cut across the granite. The rocks in the dikes are often removed by wave action, forming a slot in the rock (Figure 9–2).

The headlands are slowly cut back by wave action, forming wave-cut benches and filling in the coves with beaches of sand and gravel. The joints are enlarged by weathering and wave action to form coves and beautiful tide pools. The fir-covered mountains and rugged coast make Acadia National Park a lovely place to visit.

Sedimentary and metamorphic coasts

Sedimentary rocks are carved into steep cliffs and form some of the most spectacular coastal scenery. In contrast to the low-lying, wooded headlands and narrow bays characteristic of the igneous rock coast, the sedimentary rock coast is relatively straight and often has long, steep cliffs (Figure 9–3).

Granite consists of large, interlocking crystals of quartz and feldspar, which cling together to form a tough, resistant rock. Therefore, the granite coast erodes slowly and forms low, rolling headlands separated by deep bays. In contrast, sedimentary rocks are formed from mud, grains of sand, pebbles or cobbles, or shells of marine organisms. Mud is compressed to form

FIGURE 9–3. Tilted sedimentary rocks are eroded in a sea cliff at Joggins, Nova Scotia.
Photo by W. T. Fox.

shale, which is the weakest sedimentary rock. Sandstone and conglomerate are formed from sand and gravel that have been cemented together by silica or calcite to form fairly resistant rocks (see Chapter Five). Limestone consists of fossil shells cemented together, or lime mud, and weathers fairly rapidly. The White Cliffs of Dover, along the southern coast of England, are examples of straight limestone cliffs that were eroded by waves.

Sedimentary rocks were originally laid down as flat layers that were almost parallel to the surface of the earth. In some areas, the sedimentary layers are folded and tilted on edge by forces within the earth's crust. The folded rocks are often exposed along the coast, where they are eroded by wave action (Figure 9–3). If the folded layers of sedimentary rocks are roughly parallel to the coast, resistant layers of sandstone and conglomerate form overhanging ledges and numerous tidal pools. A wave-cut platform forms at the base of the cliff with the resistant layers extending as ribs into the surf zone at low tide.

Folded sedimentary rocks are beautifully exposed along the 80-kilometer (50-mile) stretch of the coast in the Olympic National Park in Washington. Many wave-cut features that are typical of the sedimentary rock coast are dramatically displayed along the Oregon coast (Figure 9–4).

FIGURE 9–4. Tilted sandstone is eroded into unusual shapes at the Devil's Punchbowl on the Oregon coast. *Photo by W. T. Fox.*

FIGURE 9–5. Flat, lying sedimentary rocks are eroded into sea
stacks that are called the Twelve Apostles on the Queensland
of Australia.
Courtesy of the Australian Tourist Commission.

Wave erosion proceeds in a series of stages in the weak
sedimentary rocks. When they are submerged, the waves cut a
small slot, called a nip, into the rocks. As the nip is enlarged,
the hillside is eroded back, forming a cliff. Rocks that are
eroded from the cliff are ground up by the waves and provide
angular boulders, cobbles, and pebbles along the base of the
cliff. The rocks act as grist, which is used by the waves to grind
away at the base of the cliff. Sometimes, the cliff erodes un-
evenly and large pillars of rock, called stacks, are left standing
on an abrasion platform. The Twelve Apostles on the Queensland coast of Southern Australia are beautiful examples of
stacks (Figure 9–5). Sea caves and arches are formed by uneven
erosion along the coast.

Metamorphic rocks are formed where sedimentary or igneous rocks are subjected to increased heat and, or pressure.
Where sandstone is deeply buried and intense pressure is applied during a continental collision, sandstone is metamorphosed into quartzite. Quartz grains are recrystallized into a
dense, resistant rock that stands up well under the attack of

waves. Limestone is metamorphosed into marble, which resembles a well-cemented limestone when it comes to wave resistance. Interbedded shale and sandstone subjected to heat and pressure form schist or gneiss, which are very resistant to wave erosion. The schist has tiny grains of mica glistening on the bedding surface, but shows a highly contorted, almost fluid, flow in cross-section. The gneiss has distinct light and dark bands formed by interbedded layers of quartz and dark, heavy minerals. Since gneiss has neither the weak layering of sedimentary rocks nor the network of joints found in granite, it forms a very rugged and resistant coastline.

Life along the Rocky Coast

The organisms that inhabit the rocky coast include a limited number of algae, and animals from the major invertebrate phyla. The trees, grasses, and flowering plants so common on the land are not found in the intertidal zones. Microscopic algae form a thin slime, covering the rock surfaces, and a few large, sturdy species develop holdfasts, stems or stipes, and rubbery blades in the surf zone.

The animals on the rocky coast are largely represented by four invertebrate phyla: Mollusca, Echinodermata, Arthropoda, and Cnidaria. Snails, limpets, mussels, and clams in the phylum Mollusca are well adapted to the intertidal zone. Starfish and sea urchins belong to the phylum Echinodermata. Arthropods, which are dominated by insects on land, are represented by barnacles and crabs in the intertidal zone. Corals, jellyfish, and sea anemones are members of the phylum Cnidaria.

The vertebrates are notably absent as full-time residents of the intertidal zone. However, mammals, reptiles, amphibians,

FIGURE 9–6. A sea otter eating an abalone while floating on its back in the waters of Monterey, California. *Drawing by C. W. L.*

FIGURE 9–7. The California sea lions gather in herds along
the rocky Pacific coasts as far south as the Galapagos Islands.
Drawing by C. W. L.

and birds invade the littoral zone when the tide is out, and sea
mammals and fish feed along the shore when the tide is in. Liz-
ards, raccoons, and other small animals forage among the mussels and crabs at low tide. Along the California coast, sea otters
and sea lions feed on abalone, sea urchins, and crabs (Figures
9–6 and 9–7).

It is only possible to give a brief introduction to each of
the major phyla in the following sections, and further reading
listed at the end of the chapter is recommended for more information on each of the groups.

Intertidal plants   Where the waves are not too violent, rocks in the intertidal zone
are draped with a thick layer of seaweed. Most of the plants are
different species of algae that are well adapted to the waves and
tides. The upper part of the intertidal zone is covered with a
black, crustose lichen, *Verrucaria,* which gives the appearance
of a black strip along the rocks. *Calothrix,* a simple blue-green
algae, also forms the dark slime in the black zone.

The intertidal zone is characterized by several large species of algae. On the southwest coast of England and on the
New England coast of North America, species of algae are restricted to distinct but overlapping zones (Figure 9–8). *Ulva,* or
sea lettuce, is bright green in color with an average size of 10 to

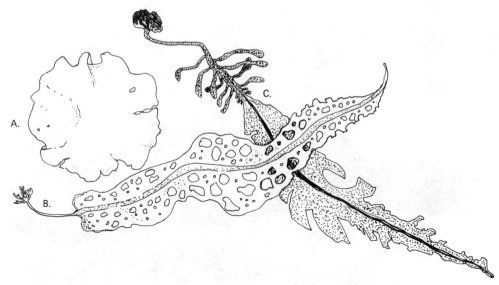

FIGURE 9–8. (A) Sea lettuce, *Ulva* (length: 100 cm); (B) sea colander, *Agarum* (length: 2 m); and (C) edible kelp, *Alaria* (length: 4 m).
Drawing by C. W. L.

30 centimeters. The individual blades of sea lettuce are lanceolate, or long and tapered, and are often ruffled or folded. The brown algae, *Ectocarpus,* is a delicate branching form with many fine hairlike strands on the branches. *Ralfsia* is the common olive-brown algae found on the rocks of the New England coast. Several species of *Alaria* with broad, flat blades occur in the lower part of the intertidal zone.

Several species of fucoids (meaning rockweed), including the channelled wrack *Pelvetia* in the upper zone, are followed by a thin zone of *Fucus spiralis.* The main intertidal zone is covered by a mixed growth of *Ascophyllum nodosum* and *Fucus vesiculosus.* A belt of the toothed wrack, *Fucus serratus,* extends into the lower part of the intertidal zone. On the New England coast, the lower part of the intertidal zone is dominated by *Chondrus crispus,* Irish moss, and dulse, *Rhodymenia* (Figure 9–9).

The lowest intertidal zone, which remains submerged for most of the tidal cycle, is the area marked by large kelp, or *Laminaria* (Figure 9–9). There are several different species of kelp with flat blades and thick, round stipes. On the California coast, the bull or ribbon kelp, *Nemocystis,* is cut into small sections and soaked to make kelp pickles. Although it grows more than 30 meters tall in kelp forests, only the stipes less than 5 meters long are used for the pickles. The stipe ends in a hollow ball or holdfast and looks like a bull whip when it is washed up on the beach. The holdfast, with its small, entangled root system, provides a refuge for many small animals.

FIGURE 9–9. (A) Kelp, *Laminaria* (length: 5 m); (B) rock-weed, *Ascophyllum* (length: .6 m); (C) Irish moss, *Chondrus* (length: 2.5 cm); and (D) dulse, *Rhodymena* (length: .5 m). *Drawing by C. W. L.*

Cnidaria—sea nettles

The Phylum Cnidaria includes three major classes: Hydrozoa—the hydroids; Scyphozoa—the common jellyfish; and Anthozoa—the sea anemones and corals. They are primitive but successful animals that are found throughout the oceans, but seem to prefer warm, shallow water. They have been referred to as the Phylum Coelenterata, which is still widely used as a synonym for Cnidaria.

The Cnidaria have two life styles, free-swimming and attached, which have resulted in two distinct body forms, the medusa or jellyfish form and the polyp or sea anemone form. The medusa, which resembles an umbrella, has a mouth facing downward, surrounded by a ring of tentacles. The polyp is like an upside-down medusa attached at its base with its mouth and tentacles facing upward. Hydrozoa exist as single or colonial polyps, or medusa. Scyphazoa spend most of their life as jellyfish, floating around the oceans. Anthozoa have only polyps and are best known for their beautiful colors and diverse shapes.

Cnidaria are armed with stinging cells, or nematocysts. Each nematocyst consists of an inverted tube, a cap, and the fluid contents, which may be toxic. Several different types of nematocysts that will stun, paralyze, or kill their enemies or prey have been described. When the nematocyst is activated, the tube turns inside out and the contents are discharged. Most nematocysts cannot injure humans, but the stinging coral, *Millepora,* and the Portuguese man-of-war, *Physalia,* can inflict painful wounds and on occasion cause death to a person. The nematocysts line the tentacles that drag behind a jellyfish or surround the mouth of a coral. Nematocyst cells are independent of the main body of a jellyfish, and even when a Portuguese man-of-war has washed up on the beach and appears dead, the nematocysts can still be released.

Hydra.   Hydrozoa are best known for the freshwater *Hydra,* which is often studied in introductory biology classes. Delicate branching hydrozoa occur as feathery white plumes and small, transparent twigs in tide pools. The colonial hydrozoan, *Plumularia,* is almost invisible in a tide pool and must be removed to be studied. The treelike hydrozoa is a host for a diverse assemblage of smaller animals, which live among its branches. The tiny animals living with the hydrozoan include round worms, sea spiders, isopods, amphipods, and other minute hydroids. When naturalists are studying life in a tide pool, the hydroids are often overlooked in favor of more obvious animals and plants.

Jellyfish.   The true jellyfish of the class Scyphozoa are not residents of the intertidal zone, but often wash up on the rocks or are seen floating near the shore. The large moon jellyfish, *Aurelia,* is about 30 centimeters wide and consists of a gelatinous dome with four simple lobes trailing from the mouth (Figure 9–10). Small jellyfish are generally harmless and are interesting to watch as they pulse through the water. Although most jellyfish won't cause any harm, it is generally not good to handle

FIGURE 9–10. The moon Jelly *Aurelia* floats through the wave (length: up to 30 cm).
Drawing by C. W. L.

them; one might receive a painful sting or, if one is sensitive, an allergenic reaction. The largest jellyfish, *Cyanae arctica*, which lives in the Arctic, weights almost a ton and has tentacles up to 30 meters long.

Sea anemones and corals. The Anthozoa, or flower animals, include the sea anemones and corals. The corals, which form a hard cup of calcium carbonate, will be further discussed in Chapter Ten.

The sea anemones are common residents of the intertidal zone and can be easily mistaken for plants. The sea anemones are attached to the rock by a basal disc, but they can move when the need arises. When the anemone is feeding, the polyp is open and tentacles extend waving into the water, gathering food toward the mouth. The tentacles have small cilia on their surface, which create gentle currents to move small food particles toward the mouth. The tentacles also have many nematocysts and adhesive glands, which aid in capturing large particles. Sea anemones can reproduce sexually or asexually by budding or cloning.

On the California coast, the rocks in the upper intertidal zone are often covered with small anemone. When the tide is out, the tentacles are contracted and the anemones look like a slimy covering on the rocks. The dull brown surface of the closed anemones is often covered with bits of rock and other debris, disguising their presence. When the tide is in, the anemones open up and extend their tentacles. The anemones are voracious predators, and individual anemones have been kept alive for up to 80 years in an aquarium.

In the lowest tidal zone on the California coast, the giant green sea anemone, *Anthopleura*, are up to 25 centimeters in diameter (Figure 9–11). These solitary forms have a uniform green color when the tentacles are open, and resemble amor-

FIGURE 9–11. The giant green sea anemone *Anthopleura* being fed upon by Wentletraps and other snails (diameter: up to 25 cm).
Drawing by C. W. L.

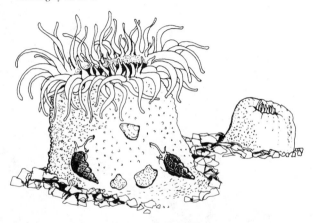

phous brown blobs when they are exposed. In contrast to the smaller anemones, the giant anemones can exist only in the well-oxygenated surf zone and cannot live where sewage or pollution are present. Several different species of anemones live in the tidal pools on the Atlantic coast, with colors varying from bright green to red or orange.

Arthropods: jointed appendages

The phylum Arthropoda must be considered the most successful animal phylum on the surface of the earth. The trilobites that ruled the Paleozoic seas became extinct about 240 million years ago. The class Chelicerata includes spiders, scorpions, and ticks, which are widespread on land. The class Manibulata, which includes the insects on land and Crustacea at sea, are the most abundant and diverse group of animals living today. Among the Crustacea, the barnacles and crabs have invaded the intertidal zone and cover much of the available rock surface in the midtoral zone.

Barnacles. Barnacles by the billions cover rocks, pilings, turtles, and other hard objects in the intertidal zone. Barnacles are unique among the arthropods in that they are attached by cement and spend the rest of their lives supported by their necks. They build small, volcanolike shells with calcareous plates that enclose their bodies (Figure 9–12). Their heads are large, but their abdomens are almost missing. The mid-section, or thorasic region, bears six pairs of legs, which are used for feeding. The outer legs are covered with small hairs, or cirri, which generate a weak current toward the mouth. Barnacles filter small planktonic organisms, but if a larger morsel stumbles within their grasp, it is passed up to the mouth by the legs.

FIGURE 9–12. (A) Acorn barnacles *Balanus* feeding just below the water surface; (B) cross-section of a barnacle, using extended appendages to set up currents that move small food toward the waiting mouth (width: 3 cm).
Drawing by C. W. L.

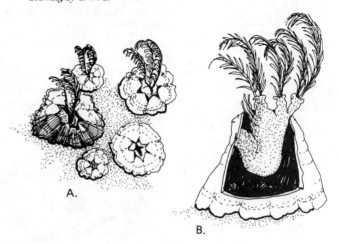

A.

B.

The barnacles have followed two evolutionary paths, which can be seen in the intertidal zone. The acorn barnacles are squat and form small cones plastered against the rocks. The goose barnacles have a fleshy stalk, known as a peduncle, which is attached to rocks, floating logs, or ships.

In the upper part of the intertidal zone, the large acorn barnacle, *Balanus,* covers the rocks with up to 4 individuals per square centimeter (Figure 9–12). When the acorn barnacles are feeding, the upper plates open and the legs extend to actively bring food to the mouth. Some species of barnacles high in the midlittoral zone are only under water for brief intervals each day and open almost immediately when they are submerged. Other species of barnacles live in the infralittoral zone and are intolerant to exposure. They are also found in tide pools and under algae.

The genus *Balanus,* which can be distinguished by its large size and its rough, white appearance, is considered a boreo-arctic form that is restricted to the cold northern waters. The small barnacle with a brown coloration and a smooth surface belongs to the genus *Chthamalus,* which is a tropical form. On many coastlines, the northern range of *Chthamalus* overlaps with the southern extent of *Balanus.* Where strong surf pounds directly on the rocks, the zone of *Balanus* expands at the expense of *Chthamalus.*

Under crowded conditions, the acorn barnacles change from flattened to towering forms. Most of the barnacles are of about the same size because a single brood is released in late spring. They grow rapidly to fill the available space and rapidly colonize any new surface that is presented, such as a piling or bottom of a ship.

The stalked or leaf barnacle, *Pollicipes,* forms dense clusters on the Pacific coast from Canada to Mexico (Figure 9–13). The leaf barnacles are sensitive to light and can move in search of prey. The goose barnacle, *Lepas,* is not a shore dweller, but can be seen on logs that drift at sea. The name for the goose barnacle comes from a zoology text written in 1596, in which the author claimed that baby geese fell feet first from the head of the goose barnacles into the sea. The delicate plates on the side of the goose barnacle also resemble a horse's hoof.

FIGURE 9–13.
The leaf barnacle *Pollicipes* cluster around a limpet and some acorn barnacles on the west coast rocky coast (approximate length: 10 cm).
Drawing by C. W. L.

Hermit crabs. Crabs are the clowns of the tidal pools. Hermit crabs, *Pagurus,* have evolved into a dependence on snail shells for their homes. Their abdomen has been converted into a hook that reaches into an empty snail shell, which is then carried on the back of the crab (Figure 9–14). As the crab grows, it must trade in its old shell for a larger model. When a hermit crab sees a better shell, it follows a ritual motion, discarding its old shell and trying on the new one. In the ritual, it first inspects the new shell, touches it, grasps it, then rotates it until the opening is in a position to be explored by the antennae. If the new

FIGURE 9–14. Hermit crabs *Pagurus* scuttle about in rocky tide pools, dragging their snail shell home behind them (length: up to 4 cm).
Drawing by C. W. L.

shell is found satisfactory, the crab moves in. If you place a small hermit crab on the palm of your hand, it will sit idle for a few moments, then bravely move out of the shell to explore your hand. Although they look aggressive, most of their diet consists of microalgae.

Several different types of small, elusive crabs also live in the tide pools and among the algae. The rock crab with a dark green or red square shell is an active scavenger among the rocks and tide pools. Its rapid sideways motion makes this small crab difficult to catch. The larger crabs, which can provide a meal, are usually found in the mud or sand bottoms of the bays and estuaries.

Mollusca— soft-bodied animals

Four major classes of Mollusca—Polyplacophora, Gastropoda, Pelecypoda, and Cephalopoda—are well adapted to the rocky coast. Each of the major classes of mollusca that inhabit the intertidal zone are adapted to different modes of life. The chitons or sea cradles clinging to the underside of rock ledges are polyplacophorans. The gastropods, which include the snails, limpets, and sea-slugs, cling to the rocks and inhabit the tide pools. The mussels, which are attached to the rock by threads, are pelecypods. Squids and octopi, which roam in and out of the littoral zone, are cephalopods.

The mollusca have five major parts in common that are expressed in different ways in the major classes. Each group has a head with sensory organs, a muscular foot for locomotion or attachment, a visceral mass containing the vital organs, a mantle that folds over the visceral mass and secretes a calcium carbonate shell, and a mantle cavity containing the gills. The foot acts as a platform holding the viscera, which is covered by the mantle and shell or forms tentacles.

In the chitons, the visceral mass is flattened and the mantle cavity is extended forward as a groove around the foot. The chiton shell is divided into eight overlapping plates.

In mussels and clams, the mantle is enlarged and hangs down in symmetrical lobes on each side of the body. The gills are enlarged within the mantle cavity. The foot of the clam becomes a blade or digging tool that extends downward between the folds of the mantle. The mantle secretes a pair of oval shells that open along a hingeline.

In snails, the visceral mass grows upward to become a visceral hump that is twisted within a spiral shell. The snails have a distinct head with extended stalks that often bear eyes. A strong foot is used for moving slowly over the rocks or holding fast in one place.

Squids and octopi also have a raised visceral hump, and the early cephalopods had a chambered shell. However, most of the living cephalopods have a reduced internal shell. Both the squid and octopus squeeze water from the mantle cavity for a jet assist in swimming. The foot in the octopus has been converted to long tentacles lined with powerful suckers. The octopus head contains a brain and large, well-developed eyes.

Snails and limpets. The snails are by far the most common and diverse mollusks on the rocky coast. Gastropod means stomach-foot, referring to the large powerful foot underlying the gut of the snail. In the upper tidal zone, the genus *Littorina*, or "shore dwellers," cling to the rocks and crevices. The *Littorina*, or periwinkles, are characterized by a dull gray shell with three or four whorls in the spiral, but some species are brightly colored or striped (Figured 9–15).

Periwinkles, like most of the gastropods, have a radula or coiled strip containing a series of teeth, which they use to scrape algae or benthic diatoms off the rocks. They also have an operculum or oval plate attached to the upper surface of the foot, which acts as a trap door when the snail pulls back into its shell. The operculum provides protection against predators, and also forms a seal to retain moisture in the gills and keep out fresh water.

When a periwinkle was fed to a sea anenome in an aquarium, the anenome swallowed the snail and spit it out 12 hours later with a polished shell. A few moments after it emerged, the periwinkle lifted its operculum and resumed its normal life. Periwinkles have been kept out of seawater for 42 days without apparent harmful effects.

The foot of the periwinkle is adapted for moving around on the rock surfaces. Glands within the foot secrete mucus, which acts as a lubricant for moving and provides suction for holding onto the rock. The foot is divided up the middle so that the right and left halves function independently. Therefore, the periwinkle is able to glide over or skate along the rock surface.

The limpets are small snails with conical shells that resemble a chinese coolie hat (Figure 9–16). Two genera of limpets, *Patella* and *Acmaea,* scour the rocks in the intertidal zone.

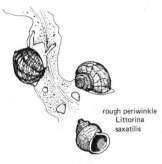

rough periwinkle
Littorina
saxatilis

FIGURE 9–15.
Periwinkles: *Littorina*, feeding on rockweed (size: about 4 cm); rough periwinkle Littorina saxatilis.
Drawing by C. W. L.

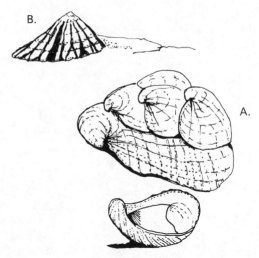

FIGURE 9–16. (A) Common slipper shells, *Crepidula*, stack one on another. The larger below is female; the smaller above are males (length: 5 cm). (B) Tortoise-shell limpet clings to rock face (length: 2.5 cm). *Drawing by C. W. L.*

The limpets roam at night or when the tide is in, and spend the daylight hours plastered against the rocks. Many limpets browse for food and return to the same spot, which becomes their home. The outer margin of the limpet shell fits snugly against the rock, making it difficult to dislodge the limpet by waves or predators. Most small limpets, from 1 to 5 centimeters across, trace a *U*-shaped path along the rocks as they forage for algae and organic detritus.

The keyhole limpet has a hole at the top of the cone with the gills suspended beneath the hole. The hole allows water to flow through the gills without lifting the margin of the shell off the rock. The largest keyhole limpet, *Megathura*, up to 18 centimeters across, lives along the coast of southern California and northern Mexico.

The abalone are large gastropods that cling to the rocks in the lowest intertidal zone. The red abalone, *Haliotis*, is sought often along the California coast. It has a broad edible foot, similar to that of the limpet. The single shell is oval and flattened with rough corrugations or growth lines along the surface. It has a low spire near the posterior end of the shell, which shows its relationship to the spiral gastropods. The outer edge of the shell is thin and sharp, and the inner edge is thickened and turned under, with a row of holes along the top. Siphons that discharge water and waste material protrude through the holes.

The abalone used to be abundant along the California coast, but their numbers decreased as they were hunted for food. The large foot must be pounded into an abalone steak and

FIGURE 9–17.
The dogwinkle, *Thais,* is a carnivorous snail that feeds on mussels and barnacles (length: 2.5 cm).
Drawing by C. W. L.

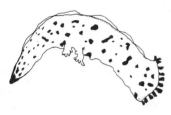

FIGURE 9–18.
Nudibranch, *Triopha carpenteri,* occurs in the high-tide pools near Monterey, California (length: about 2.5 cm).
Drawing by C. W. L.

let stand for 24 hours before it can be eaten. The shells are also quite attractive, with the shimmering iridescence or mother-of-pearl on the inner lining. The outer surface of the shell is often covered with barnacles or other marine organisms.

Although most snails on the rocky coast are herbivores, feeding on various types of algae, the dogwinkle, *Thais,* and the moon snail, *Policines,* are carnivores. The radula with the aid of secreted enzymes is used to drill holes through shells. The dogwinkle has a spiral shell sometimes bearing dark strips parallel to the opening (Figure 9–17). The dogwinkle can often be found feeding on mussels or barnacles in the tide pools or along the rocks. The moon snail has a heavy, smooth shell and may be as large as an apple. The large beveled holes found in many clam shells on the beach are probably the result of moon snail boring. The moon snails themselves can also provide the base for a hearty sea food chowder.

The nudibranch or sea slugs are gastropods without shells and are also predators in the intertidal zone. The nudibranchs have brightly colored serata, or hairlike appendages, surrounding the anus or scattered along the back (Figure 9–18). In the tide pools of California, the yellow nudibranch, *Anisodoris,* has black tubercles along its back. The nudibranchs appear defenseless, but they must be unappetizing because they are generally not eaten by the others in the tide pools.

The slipper shell or boat shell has a slight twist at the apex of the shell, and a small platform on the inside. The Pacific coast species, *Crepidula adunca,* is known as the boat shell. The Atlantic coast species, *Crepidula fornicata,* form long chains containing up to thirteen individuals (Figure 9–16). The lowest individual is a large old female; the middle one, hermophroditic; and the smaller, upper ones are males. The individuals are first male, then bisexual for a time, and finally female. The old individuals are permanently attached to their support by calcareous secretions, and the younger ones are partially interlocked by projections and grooves.

Mussels, oysters, and clams. Pelecypods, or bivalves, in the intertidal zone are represented by the mussels, oysters, clams, and scallops. The mussel, *Mytilus,* lives firmly attached to the rocks by strong elastic threads forming the byssus (Figure 9–19). Mussels are usually found in dense clusters in the intertidal zone below or overlapping with the barnacles. Mussels are able to move a short distance by breaking some threads in the byssus and attaching new ones to the rocks.

The mussel shell is elongated with rough concentric growth lines parallel to the outer margin. The outside of the shell is black, brown, or blue, giving it the familiar name, the blue mussel. The inside of the shell is a lighter shade than the outside and has a pearly luster.

Mussels are efficient suspension feeders, living on phytoplankton and floating detritus in the surf zone. Although the

243

FIGURE 9–19.
California mussels, *Mytilus
californianus*, with acorn
barnacles and limpets growing on
shells. Byssal threads attach the
mussels firmly to their rocky beds
(length: 10 to 15 cm).
Drawing by C. W. L.

mussels are rich in vitamins, their filtering habit makes them collectors of the toxic dinoflagellate, *Gonyaulux,* which causes the red tide. Care must be taken in eating mussels during the summer months when the red tide becomes abundant. In polluted areas, mussels should also be avoided because their livers concentrate harmful nitrates from the water.

Oysters and clams are generally found in protected areas and are not abundant on the exposed rocky coast. Oysters generally occur near the 1-meter tide level on the Pacific coast, and are restricted to bays and estuaries on the Atlantic coast. The clams are burrowing animals, abundant in the sand flats and pocket beaches along the rocky coast.

The fan scallop shell was used as a Crusaders' badge in the Middle Ages and today is the logo for a major oil company. Most scallops are free-swimming clams that propel themselves by rapidly opening and closing their shells. Although most scallops live in the sheltered bays, swimming among eel grass, the rock scallop, *Himmites,* is attached to the rocks in the lower part of the rocky coast and only swims free when it is a juvenile.

Octopi and squids. Cephalopods include octopi with eight arms, squids with ten arms, and the chambered Naulitus with a shell and more than ninety tentacles. Octopi live in the low intertidal zone, hiding in the rock crevices and protected areas along the bottom. The arms of the octopi contain flattened, disc-shaped suckers. Some species of octopi are mildly venomous and have been known to inflict painful wounds; however, they are more commonly shy and retiring. They eject a dark fluid blob to simulate their rotund body when they make their escape. Although they can crawl on their arms and use the webbed skin between the tentacles for swimming, their most efficient mode of locomotion is jet propulsion. The octopus pumps water out of the mantle cavity by alternate contractions of the circular and radial muscles and moves through the water in a series of short, rapid surges.

Squids swim in schools near the coast, with larger members of the group occupying the ocean deeps. The giant squid, *Architeuthis,* may reach 18 meters or more in length and has given rise to fantastic tales of sea monsters. However, the common squid, *Loligo,* is less than a meter, usually only 20 to 30 centimeters long. The common squid, called calimari, is considered a tasty dish in Italy and along the California coast.

The cephalopods are worth considering a little further, because they have the most highly developed nervous systems in the invertebrate world. Their advanced nervous system is matched by their complex behavior patterns. The brain of the octopus has several lobes with distinct centers for controlling specific activities. Giant nerve cells, or neurons, up to a meter long, extracted from squids have been used for research in neuron physiology. Although the shelled cephalopods reigned supreme 500 million years ago in the Paleozoic Era, they decreased in importance about 240 million years ago in Mesozoic and Cenozoic Eras. It is interesting to speculate where the cephalopods would be today if they had developed the air-breathing capability of the gastropods to match their advanced wits. Since the mantle cavity that contained the gills also developed into a means of locomotion under water, the cephalopods are tied to the marine realm.

Chitons. The chitons occur in all climates and in the depths of the sea down to about 4000 meters. Most chitons, which are also known as sea cradles, have eight overlapping plates arranged in a row along their back. The eight plates are held in place by a muscular ring or girdle around their margin (Figure 9–20). The typical California chiton, *Tonicella,* is about 3 centimeters long and spends most of its time on the underside of rocks. Like the limpets, some chitons have a home spot on the rock to which they return after foraging for algae or seaweed. The giant gumboot chiton, *Cryptochiton,* which roams in the low intertidal zone on the Pacific coast, is almost 30 centimeters long. In the gumboot chiton, the muscular girdle completely covers the plates, giving it the appearance of a massive blob. All species of chitons have a smooth foot that is edible but hardly worth the effort. It must be pounded and tenderized before it can be fried or put in chowder.

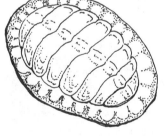

FIGURE 9–20.
The red chiton, *Ischnuchiton ruber,* has a shell consisting of eight overlapping plates (length: about 7 cm).
Drawing by C. W. L.

Echinodermata: spiny-skinned animals

Many of the most interesting and attractive animals in the intertidal zone belong to the Phylum Echinodermata. The word *echinoderm* means "spiny skin" for the many different types of spines or projections that they use for protection and locomotion.

There are four distinct classes of echinoderms: Asteroidea—commonly called starfish, but sea stars may be a better name because they are not related to fish; Ophiuroidea—brittle stars, or serpent stars; Echinoidea—sea urchins; and Holothuroidea—sea cucumbers. The echinoderms cannot regulate their salt water content and constantly circulate sea water through their bodies. Therefore, they have never been able to move into the fresh water habitats and are exclusively marine.

The echinoderms have a radial symmetry that has been superimposed on a primitive bilateral symmetry. In a radial symmetry, the arms radiate out from the center like spokes on a wheel. Bilateral symmetry indicates that the two sides of an organism are mirror images of each other, with an equal number of arms or legs on each side. Mammals have a bilateral symmetry, while many flowers have a radial symmetry.

Most echinoderms begin life as bilaterally symmetrical organisms and metamorphose into radial forms. Many of the echinoderms have a fivefold symmetry as can be seen from the five arms of a starfish or five petals on the surface of a sand dollar. Several of the modern free-moving forms have superimposed a secondary bilateral symmetry on the earlier radial symmetry.

The echinoderms have developed a unique salt water circulation, called the water vascular system, for respiration, movement, and sensory perception. The water vascular system consists of a ring canal that encircles the mouth and radial canals that extend into the arms. The radial canals contain openings for the tube-feet, or podia. Sea water is pumped into the system through the madrepore near the mouth, which acts as a sieve. The water is circulated through the ring canal and radial canals, providing hydrostatic pressure or suction at the ends of the tube-feet.

The sea star can pump water into the tube-feet and actually walk along a rock surface, or it can expel water, forming a strong suction that holds the animal to the surface of the rock. If the sea star is pulled off the rock too quickly, the tube-feet are left behind sticking to the rock. The tube-feet are also used for grasping and opening clam shells, which provide food for the sea star. The mouth is usually located at the center of the radiating arms on the lower side, with the anus on the upper surface.

The echinoderms are noted for their body-wall appendages, which are distributed on the outer surface. The specialized appendages include spines, pedicellaria, and podia, which are emphasized to a different degree in each group. The spines, which are best developed on the sea urchins, are specialized for diverse functions, from walking to digging and self-defense.

The pedicellaria are tiny jaws that are either stalked, attached directly to the outer surface, or sunk in depressions beneath the surface. The pedicellaria snap at intruders protecting the delicate tube-feet and also remove sand grains or other particles from the surface. If you place your hand on the upper surface of a sea star, you will often feel a tingling sensation as the pedicellaria snap at your fingers.

The different classes of echinoderms can be distinguished by the orientation of the body and the distribution of the arms. The Asteroidea, or sea stars, are oriented with their mouths down and have open grooves extending along the broad arms. The Ophiuroidea, or brittle stars, also lie with their mouths down, but they have narrow, flexible arms. The Echinoidea, including the sea urchins and sand dollars, normally lie with their mouths down and have their arms wrapped upward around the sides of the body. When the spines are removed from a sea urchin, the grooves are clearly visible, and the shell looks like a sea star with the arms gathered upward and tied at the top. The Holothurians, or sea cucumbers, also have grooves that resemble arms, but they are elongated along the oral-aboral axis, from the mouth to the anus.

Sea stars. The Asteroidea, or sea stars, are often found in the mid and lower intertidal zones, where they feed on almost all other intertidal animals. At low tide, the sea stars can be found clinging to the undersides of rocks or occasionally having a meal on a mussel or snail. The Ochre star, *Pisaster,* is the common sea star found on the rocky shore of California. Other species of *Pisaster* are found along most of the rocky coasts around the world (Figure 9–21A). *Pisaster* usually has short spines and its color varies from orange through brown or purple. The upper-body wall is protected by pedicellaria, with tube-feet or podia on the lowe side.

Pyncopodia is the many-rayed sunflower star that occurs low in the intertidal zone (Figure 9–21B). The sunflower stars are the largest known sea stars, up to 60 centimeters across. Although they can move their arms to right themselves, the tube-feet act as legs and allow the sea star to walk with ease on a solid surface or on sand. The sunflower star begins life with six rays, which are increased to more than twenty when the animal reaches maturity. It generally has about twenty-one arms and is sometimes called the twenty-one pointer. The spiny sea urchin is one of the staples of the sunflower star diet.

Although some sea stars swallow food whole, others actually turn their stomachs inside out through their mouth and press it against the organism being consumed. Enzymes, which soften the food and predigest it before it is swallowed, are secreted from the stomach. When a sea star struggles with a mussel, it grasps the mussel with its podia and applies pressure to open the shell. When the mussel relaxes, a lobe of the stomach can enter an opening of no more than 0.1 millimeter and start to digest the mussel. When the mussel is weakened by being eaten alive, the shell opens farther and the sea star completes its meal.

Some sea stars have the unusual ability to regenerate severed arms, which is known as autotomy. The sea star, *Linckia,* one of the extreme examples of autotomy, lives on the southern

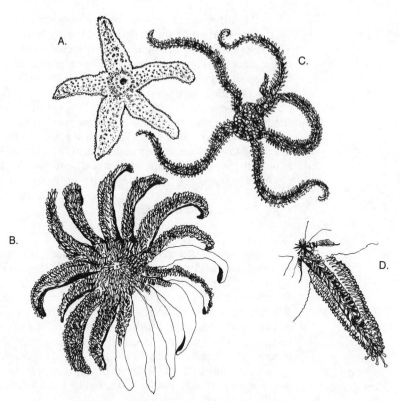

FIGURE 9–21. Different types of sea stars. A. An asterid sea star, *Asterias forbesii* (length: up to 12 cm). B. The under (oral) side of the sunflower star, *Pycnopodia helianthoides* (length: about 15 to 20 cm). C. Daisy brittle star, *Ophiopholis aculeata* (arms: about 10 cm). D. Underside of a starfish, showing tube feet on a single arm and the central mouth opening.
Drawing by C. W. L.

California coast. While the normal number of rays on *Linckia* is five, specimens have been studied with anywhere from four to nine rays. *Linckia* appears to break off arms at will, and will regenerate a new arm where the old one was destroyed. In one instance, a specimen of *Linckia* was seen with the main portion of the body fixed and passive, while one arm walked slowly away at right angles to the body, twisted, and broke loose from the rest of the body. Rays that are cut from the body will generate new discs, mouths, and rays in about 6 months. Attempts to eliminate sea stars by cutting them in two and throwing the pieces back into the sea are futile when they can regenerate a new body from a disposed arm.

Brittle stars. The brittle stars in the Class Ophiuroidea are often fairly abundant in the intertidal zone, but they remain hidden under plants or rocks, and are not obvious to the casual observer. The tiny stars, *Amphioplus,* feed on the microscopic algae in the layer of slime on the rocks. The larger star, *Ophiopholis* (Figure 9–21C), is a sand-colored brittle star with a disc about 2 centimeters across and arms that extend 10 centimeters. The jointed arms and distinct central disc make the brittle stars a unique group of echinoderms. They are often abundant in the deep sea, where they are not disturbed by waves or currents.

Sea urchins.   The sea urchins in the Class Echinoidea depend on spines for locomotion and defense. The sea urchins encountered in the intertidal zone are regular echinoids with a nearly spherical shell (Figure 9–22). The irregular echinoids with a bilateral symmetry include the sand dollar, with a disc-like shape, living just below the surface of the sand, and the heart urchins, with a more bulbous form, which burrow in the soft mud. Sea urchins are voracious eaters in the subtidal zone and have been known to wipe out a whole bed of kelp on the Pacific coast.

The common west coast sea urchin, *Strongylocentrotus,* has a diameter to the tips of the spines that can exceed 20 centimeters. Many sea urchins have spines of different lengths, from long needles to short hairlike coverings. Long spines are more effective than the short ones for movement and protection, and can be moved individually or as a group. Each spine sits on a knob that acts as a ball and socket joint. When alarmed, the sea urchin can point its spines in the direction of the danger. If they are in a crevice or hole, they use the spines to wedge them-

FIGURE 9–22. Sea urchin, *Echinus esculentus,* with tube feet to help it move and underside feeding mechanism, Aristotle's lantern, exposed. *Drawing by C. W. L.*

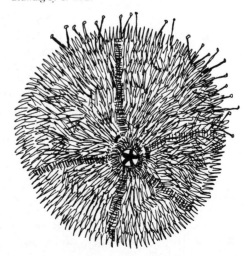

FIGURE 9–23.
Orange-footed sea cucumber, *Cucumaria frondosa*, with orange tube feet on a black body. Abundant in rock crevices in cold tide pools from Cape Cod to the Arctic. Said to be delicious boiled, but its appearance causes little interest in its edibility (length: up to 25 cm).
Drawing by C. W. L.

selves tightly against the sides. Many sea urchins are able to carve small holes in soft rock such as limestone, which provides a home. As sea urchins grow in size, they expand the hole to fit the length of the spines. Some large holes have been occupied by several generations of sea urchins.

The sea urchins also have well-developed pedicellaria, which they use for cleaning and protection. The urchins can rapidly remove a layer of sand by the combined efforts of the movable spines and the pedicellaria. Some of the pedicellaria can inflict a serious wound by injecting venom in a cut opened by a spine. Care must be taken when walking on the rocks in the intertidal zone. It is always a good idea to wear tennis shoes or boots to prevent injury from sea urchins' spines and pedicellaria.

The sea urchins have a distinctive jaw mechanism that is called Aristotle's lantern (Figure 9–22). It is a complex structure containing small bones, muscles, and teeth. It is possible for the sea urchins to extrude the lantern from their mouths and chop up large pieces of sea lettuce or other algae.

Sea cucumbers. The sea cucumbers in the class Holothuroidea are present in the littoral zone, but they are not too abundant. In the ocean at depths of 4 to 6000 meters, the sea cucumbers are the most abundant large organisms. In the intertidal zone, the genus *Cucumaria* is present as a small black animal about 3 centimeters long, and as a larger species with bright orange tentacles protruding from a dark orange body (Figure 9–23). Although there may be some confusion between sea slugs, or nudibranchs, and sea cucumbers, the sea slugs are mollusca with a well-developed foot, while the sea cucumbers lack a foot but usually have tentacles around their mouths. Most of the sea cucumbers are about the size of medium pickles, but the group ranges from tiny gherkins to majestic watermelons.

Tidal Zones on Rocky Coasts

Tides along the rocky coast are influential in the zonation of organisms. Tides are important in determining the amount of time that organisms are submerged or exposed during each tidal cycle and the length of maximum exposure or submergence between spring tides. A tidal exposure chart for an area shows the percentage of time that organisms at a particular level are exposed.

The number of hours or days of continuous submersion or emersion is important for many organisms. The zone between the lowest high water neap tide (L.H.W.N.) and the highest low water neap tide (H.L.W.N) is covered and uncovered by water during each tidal cycle. The lower or higher zones are subjected to longer intervals of submergence or emergence, ranging from a few days to several weeks. Some of the periwinkles on the high intertidal zones near the extreme high water spring tides

(E.H.W.S.) are only submerged for a brief interval at high spring tides and should probably be considered terrestrial organisms. On the other hand, sea weeds in the lower intertidal zone are submerged most of the time and are only exposed briefly at low spring tide.

Effects
of submersion
and exposure

The alternation of periods of submersion and emersion has both beneficial and damaging effects for the organisms in the intertidal zone. When the tide is in, animals and plants are bathed in water with a uniform temperature and salinity. The surging waves supply dissolved oxygen for respiration, along with nutrients and microorganisms for food. The swirling water also removes excess carbon dioxide and waste products. It is also easier for many sluggish organisms, such as limpets and snails, to move when they are under water. On the other hand, when the tide is in, predators from the sea including fish and marine mammals have access to the intertidal zone. Large waves during storms also have the power to dislodge some organisms. However, most of the essential biological activity in the intertidal zone is carried out when the tide is in.

When the tide is out, many organisms are subjected to their most difficult periods of stress. Desiccation, or loss of moisture, is the most critical problem faced by organisms in the intertidal zone. Many organisms have devised methods for trapping small amounts of sea water that can sustain them until the tide returns.

In tide pools, there are rapid fluctuations in salinity. Tide pools are flooded with fresh water during heavy rain or increase in salt during hot, dry intervals where there is high evaporation. The temperature in shallow tide polls and along exposed cliffs also varies considerably depending on the weather and the season of the year. Organisms that are exposed to the air for a long period of time are also adversely affected directly by the sun. When the predators of the sea move out with the falling tide, their place is taken by birds and small animals moving down from the land.

Intertidal zonation:
Stephenson

Distinct zones of plants and animals are obvious along most rocky coasts. Although the species are different depending on the type of coast, tide range, wave exposure, and climate, the same general zones can be recognized throughout the world. Although zones are given distinct names along different coasts, based on the dominant species of plants or animals, the major zones extend from the tropics to the arctic and are found in the Atlantic, Pacific, and Indian Oceans.

The most widely accepted scheme of coastal zonation was developed by Alan Stephenson in several research papers on intertidal life. After his death in 1961, his wife, Anne Stephenson, completed their book, *Life between Tidemarks on Rocky Shores,* which was published in 1972. From 1918 through 1961, the Stephensons studied intertidal life on the Great Barrier Reef in Australia, New Zealand, South Africa, Bermuda, the Atlantic coast of North America, the Pacific coasts of North America and Chile, and the British Isles. From their worldwide perspective, a universal scheme of zonation between tidemarks emerged. The scheme was first proposed in 1949, and has since been applied to rocky coasts throughout the world. The zonation is based on the distribution of plants and animals, and not on precise tide levels (Figure 9–24).

In the Stephenson zones, the littoral zone, or intertidal zone, extends from extreme high water of spring tides (E.H.W.S.) to extreme low water of spring tides (E.L.W.S.) (Figure 9–24). The littoral zone has been subdivided into three parts—supralittoral fringe, which also includes the splash zone above extreme high water of spring tides (E.H.W.S.); the midlittoral zone; and infralittoral fringe. The supralittoral zone is the coastal region above the splash zone, which may extend several kilometers inland. The supralittoral fringe, which is often called *Littorina* zone, is the upper boundary between land and sea. The midlittoral zone, also referred to as the barnacle or the eulittoral zone, extends from the upper limit of the laminarians to the upper limit of barnacles. The lowest zone, called the infralittoral fringe, extends downward from the upper limit of laminarians to extreme low water of spring tides (E.L.W.S.). The infralittoral zone is submerged below sea level at all times. The infralittoral fringe is referred to as sublittoral by many people, but the Stephensons chose infralittoral to avoid the possible confusion of sublittoral with supralittoral.

Although it may seem immediately obvious to propose an intertidal zonation scheme based on tide levels instead of organism distribution, a zonation based on tide levels is more difficult to apply in the field, and would not have the universal application of the Stephensons' zones. High and low tide levels change daily from spring to neap tides, and seasonally as the tidal range varies throughout the year. The limits of attached organisms, such as barnacles and laminarians, remain relatively constant throughout the year, and therefore provide useful boundaries for intertidal zonation.

The Stephenson zones of the rocky coast are based on "widespread" features that characterize the zones instead of on "universal" boundaries. The Stephensons recognized that some features may be poorly developed or missing on some coasts, but the overall scheme still carries through.

The supralittoral fringe near high water level is an arid zone that is affected by spray, but it is covered by waves only during high spring tides or during storms. Small snails of the species, *Littorina,* which are called periwinkles, congregate in the supralittoral fringe, where they graze on algae. Along the Pacific coast, the surface of the rock in the supralittoral fringe is commonly covered and blackened by encrusting algae, *Por-*

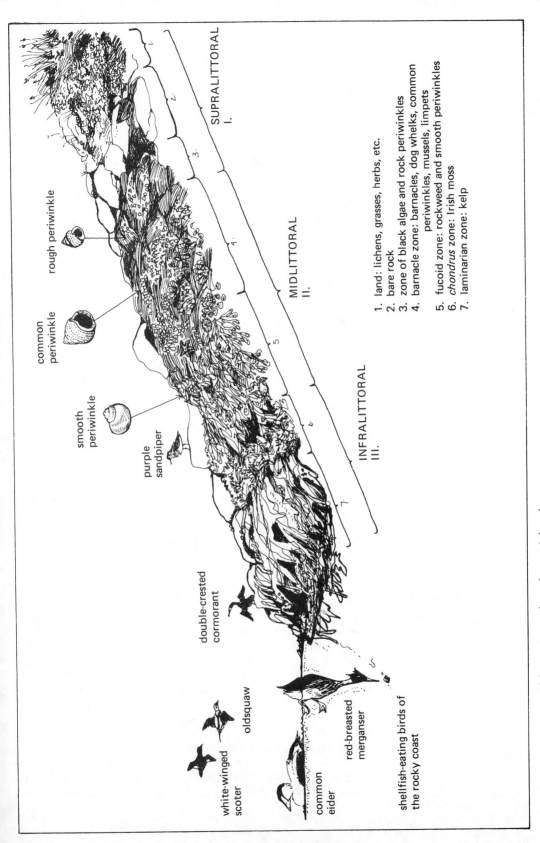

FIGURE 9–24. Intertidal zonation on the rock coast is based on the distribution of plants and animals. *Drawing by C. W. L.*

phyra, and lichens, *Verrucaria* (Figure 9–24). The black zone above high tide occurs either as a continuous band or as separate patches, but it is almost worldwide. The slippery surface on the black zone has sent many a coastal ecologist sprawling on the rocks.

The midlittoral zone is covered and uncovered by the tides at least once, and usually twice, each day. The midlittoral zone is often covered by closely spaced acorn barnacles and mussels (Figure 9–24). The rocks are covered by a thin layer of blue-green algae. Many large varieties of algae, including sea cabbage, *Hedophyllum,* are draped over the rock surfaces. Small animals, including limpets, chitons, star fish, and sea urchins, live among the primitive seaweeds and feed when the tide is in.

The lowest zone, called the infralittoral fringe, is exposed only at low spring tides. Several types of seaweed, called laminaria, live from temperate to cold regions, with dense populations of coral in the tropics. Large kelp beds live in the cold water off the Pacific coast.

The populations in the infralittoral zone are extremely variable and usually quite rich. The lower limit of organisms in the infralittoral zone is often controlled by predators. Along the coast of Washington and California, upwelling of nutrient-rich waters produces a rich crop of phytoplankton and zooplankton, which provide food for the infralittoral zone.

Calcareous algae cover the rocks in many areas, forming continuous pink sheets. The calcareous algae are included under the general heading of lithothamnia, which forms a distinct ridge in the coral reefs of the Pacific, and a widespread encrustation on the rocky shores of the rest of the world.

Waves have a significant influence on the extent and type of organisms in the intertidal zone. In protected areas that are not exposed to waves, the vertical extent of each zone is quite small. On wave-exposed coasts, each zone is broader and includes more robust species. Some species thrive on well-oxygenated water in the surf zone, while others prefer the protected areas out of the reach of the waves.

Community interactions

Interactions among organisms play an important role in the zonation of intertidal organisms on the rocky coast. The acorn barnacles, *Chthamalus* and *Balanus,* provide examples of the effects of competition and desiccation on zonation. *Chthamalus* is the southern species, ranging from the Shetland Islands, Scotland, to the Mediterranean. The northern species, *Balanus,* ranges from the Arctic to southern Spain. In areas of overlap, *Chthamalus* is restricted to the upper midlittoral zone. Therefore, it appears that *Chthamalus* is limited to the upper zone by interspecific competition with *Balanus. Balanus,* which has a faster growth rate and greater population density, eliminated most of the *Chthamalus* in the low lower zone by crowding.

Physical factors determined the upper limits of the acorn barnacles. *Balanus* has a lower tolerance to heating and desiccation than *Chthamalus.* The survival of *Chthamalus* in the upper zone is a result of its greater tolerance to desiccation. Therefore, in space-limiting systems, the lower limit of a species such as *Chthamalus* may be determined by interspecific competition, while the upper limit of *Balanus* is determined by its susceptibility to physiological shock, such as thermal or desiccation stress.

The zonation of seaweeds on the New England coast is also affected by interspecific competition and desiccation (Figure 9–25). *Fucus vesiculosus* is generally restricted to the upper part of a midlittoral zone by competition with Irish moss, *Chondus crispus,* which is abundant in the lower part of the zone. The upper limit of Irish moss is determined by desiccation, which leaves space available for *Fucus* in the upper part of the zone. Irish moss is harvested commercially as a source of agar and gelatine, which are used in toothpaste, cosmetics, lotions, and ice cream. When Irish moss is harvested, *Fucus* often becomes established in dense stands in the lower zone.

Local species diversity in the intertidal zone is also related to the efficiency of predators. Where predators are removed or limited, one of the prey species may monopolize some important limiting environmental requisite. In the marine intertidal,

FIGURE 9–25. Intertidal zones are exposed at low spring tide on Cape Ann, Massachusetts.
Photo by W. T. Fox.

the limiting requisite is usually space. Species that prevent monopolization by consuming prey species are known as keystone species. Keystone species may be carnivores or herbivores, in which the competing prey species are plants. For example, the sea star *Pisaster,* the top carnivore on the rocky intertidal zone of the Pacific coast of United States, consumes snails, barnacles, and chitons, but prefers mussels. The removal of the keystone species, *Pisaster,* accelerates the dominance of the mussel, *Mytilus.*

Summary

The character of the rocky coast is determined by the force of the waves, range of the tides, and strength of the rocks. Large storms on the Pacific coast generally produce larger waves than the Atlantic storms. The Pacific coast has a mixed tide with a high spring tidal range, while the Atlantic coast has a semi-diurnal tide with a lower range. On the Atlantic coast of North America, New England and the Maritime provinces of Canada are predominantly granitic and metamorphic rocks, with some sedimentary rocks. The rocks exposed along the Pacific northwest and California are mainly metamorphic and sedimentary, with some volcanic rocks.

Life on the rocky coasts is controlled by tidal range and wave exposure. The length of time of submergence and exposure limits the distribution of organisms. The supralittoral fringe is the highest intertidal zone and is marked by a band of black lichen that is grazed upon by periwinkle snails. The midlittoral zone, which is exposed and submerged twice each day, is dominated by mussels and barnacles. The infralittoral fringe is exposed during low spring tides and contains large rockweeds and oarweeds known as laminarians.

The major invertebrate phyla are represented in the intertidal zone. The Mollusca include snails, clams, mussels, chitons, and octopi. The Echinoderms, including sea stars, sea urchins, and sea cucumbers, roam about on the rocks. The barnacles and crabs are Arthropods related to the insects on dry land. The jellyfish and sea anemones are related to the corals in the phylum Cnidaria.

Selected Readings

AMOS, W. H. *Life on the Seashore.* New York: McGraw-Hill, Inc., 1966. Excellent photographs of different seashore habitats.

BERRILL, M. AND D. BERRILL. *A Sierra Club Naturalist's Guide to the North Atlantic coast—Cape Cod to Newfoundland.* San Francisco: Sierra Club Books, 1981. A useful pocket guide to seashore life in the northeast.

BERRILL, N. J. *The Living Tide.* New York: Dodd, Mead and Co., 1953. An interesting discussion of marine life on the east coast.

BERRILL, N. J. AND J. BERRILL. *One Thousand and One Questions Answered about the Seashore.* New York: Dover Publications, 1976.

Everything you wanted to know and were afraid to ask about the coast.

The Audubon Society Book of Marine Wildlife. New York: Harry Abrams, 1980. A stunning set of photographs of marine life.

CAREFOOT, T. *Pacific Seashores.* Seattle: University of Washington Press, 1977. A well-illustrated guide and discussion of the ecology of the Pacific coast.

CHAPMAN, C. A. *The Geology of Acadia National Park.* New Haven, Conn.: The Chatham Press, 1962. A guidebook for the general public to the geologic and coastal features of the park.

DAWSON, E. Y. *Marine Botany, An Introduction.* New York: Holt, Rinehart, Winston, 1966. A college-level botany text with complete coverage of the intertidal algae.

FURLONG, M. AND V. PILL. *Edible? Incredible?* Tacoma, Wash.: Erco, Inc., 1973. A guide to seashore life of the Pacific northwest, with instruction on how to cook it and what parts to eat.

HEDGPETH, J. W. *Introduction to Seashore Life of the San Francisco Bay Region and the Coast of Northern California.* Berkeley, Calif.: University of California Press, 1964. A pocket guide to seashore life on the Pacific coast.

HILLSON, C. J. *Seaweeds, A Color-coded, Illustrated Guide to Common Marine Plants of the East Coast of United States.* University Park, Penn.: The Pennsylvania State University Press, 1977. A complete guide to the algae in the intertidal zone.

LEWIS, J. R. *The Ecology of Rocky Shores.* London: The English Universities Press, 1972. A good introduction to the concepts of coastal ecology, with examples from the British coast.

MEGLITSCH, P. *Invertebrate Zoology* (2nd ed.). New York: Oxford University Press, 1972. An excellent college-level text on the coastal invertebrates.

MILLER, D. *The Maine Coast, A Nature Lovers Guide.* Charlotte, N.C.: East Woods Press Books, 1979. A naturalist's book on marine habitats and organisms on the Maine coast.

MOORE, H. B. *Marine Ecology.* New York: John Wiley and Sons, 1958. A college-level text on marine ecology.

NEWELL, R. C. *Biology of Intertidal Animals* (3rd ed.). New York: American Elsevier Publishing Co., 1979. A standard text on intertidal animals.

RICKETTS, E. F., J. CALVIN, AND J. W. HEDGPETH. *Between Pacific Tides* (4th ed.). Stanford, Calif.: Stanford University Press, 1968. A comprehensive guide to the Pacific coast, which is a pleasure to read.

ROBBINS, S. F. AND C. M. YENTSCH. *The Sea is All About Us.* Salem, Mass.: Peabody Museum of Salem and Cape Ann Society of Marine Science, 1973. A local guide to the algae, animals, and fish of Cape Ann, Massachusetts. Good illustrations that capture the New England coast.

STEPHENSON, T. A. AND A. STEPHENSON. *Life Between Tidemarks on Rocky Shores.* San Francisco, Calif.: W. H. Freeman, 1972. A classic book dealing with intertidal zonation on the world's coasts.

SOUTHWARD, A. J. *Life on the Sea-Shore.* Cambridge, Mass.: Harvard University Press, 1965. A handy pocket guide to animals and plants along the Atlantic coast.

SUMICH, I. L. *Biology of Marine Life.* Dubuque, Iowa: William C. Brown Co., 1976. An introductory text for marine biology.

TAIT, R. V. AND R. S. DeSANTO. *Elements of Marine Ecology* (2nd ed.). New York: Springer-Verlag, 1972. Good section on coastal ecology.

YONGE, C. M. *The Sea Shore.* London: Collins-World, 1963. An excellent introduction to seashore life, by one of the founders of modern marine ecology.

ZOTTOLI, R. *Introduction to Marine Environments.* St. Louis, Mo.: C. V. Mosby, 1976. An unusual paperback text on coastal environments.

Corals of Atlantic and Pacific reefs

TEN

Coral reef ecology

Introduction to Coral Reefs

The mention of a coral reef conjures up images of a tropical island with coconut palms, exotic flowering plants, and brightly colored birds. Actually, the islands associated with reefs are usually quite small and rather desolate with a few trees and scrubby plants. Their beauty lies hidden beneath the surf, where waves are surging among the coral heads (Figure 10–1). Colonial corals and calcareous algae build massive reefs that project into the surf. During the day, coral polyps retract into small cups within the reef, while algae are using sunlight to grow. At night, polyps extend their tentacles and capture microorganisms from the water. Schools of multicolored fish swim among the coral heads, with moray eels and sea urchins living in the cracks and crevices within the reef.

A coral reef is a ridge or mound of limestone that was formed by the action of corals and other lime-secreting organisms. Coral reefs are limited to the tropics, where they thrive in warm, clear, shallow, sunlit waters with near normal salinity. The coral reefs are generally found between 30° north and 30° south latitude. Reef-building corals grow best where the average water temperature is between 23 and 25 degrees Centigrade (73 to 77 degrees Fahrenheit). Although corals have been found living at depths greater than 60 meters (200 feet), they rarely build reefs at depths greater than 50 meters (160 feet) and most reefs are confined to depths less than 30 meters (100 feet).

Reef corals exist in a symbiotic relationship with microscopic yellow algae known as zooxanthellae. The zooxanthellae live within the outer flesh of the coral polyps and require light to carry out photosynthesis. Corals that do not congregate in reefs but live a solitary existence generally do not contain zooxanthellae. Most solitary corals live in shallow water, but they are also found several hundred meters below the surface in Norwegian fjords, and beneath the ice in the Arctic.

Reef-building Organisms

Large coral reefs are created by small coral polyps and calcareous algae. The coral polyps secrete an external skeleton composed of calcium carbonate, which becomes the framework of the coral reef. The branching corals look like underwater forests that project upward into the breaking waves. The brain corals construct massive boulder-shaped structures that provide support and protection for the tiny polyps.

Calcareous red algae live among the corals in shallow water. The algae secrete thin layers of limestone that are draped over rocks and corals. The calcareous algae also build mounds of limestone with concentric layers shaped like cabbage heads. Some of the algal heads form branches that look like corals.

Corals

Corals are primitive marine animals closely related to jellyfish and sea anemones in the phylum Cnidaria. The body of the coral known as a polyp is shaped like a cup with a mouth opening at the top (Figure 10–2). The mouth is surrounded by a ring

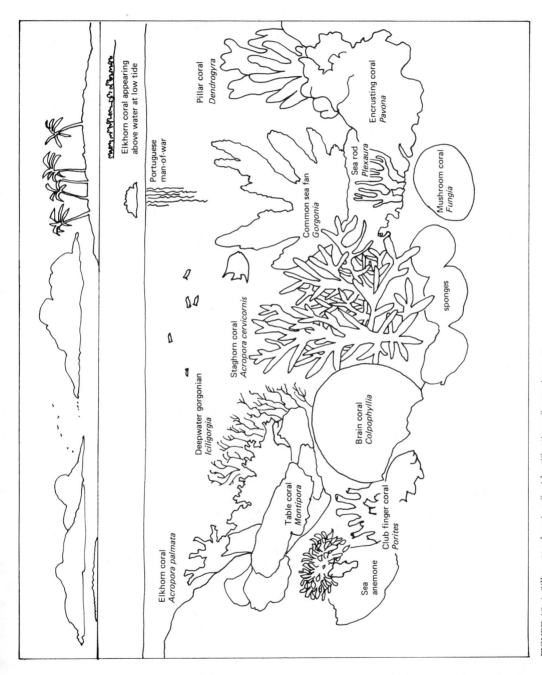

FIGURE 10–1. Silhouette drawing for identification of organisms on coral reef assemblages of Atlantic and Pacific corals. Drawing by C. W. L.

of tentacles that are used to gather food. The coral is a predator that feeds on tiny zooplankton, drifting with the waves and currents. The tentacles are covered with stinging cells and slimy mucus that traps the zooplankton like flies on flypaper. The mucus is carried to the mouth and swallowed with the entrapped microanimals. The mucus is also used to ensnare small grains of silt and sand before they clog up the polyp.

The coral polyp sits inside a small limestone cup called a corallite, which provides both support and protection. The cup, composed of a calcium carbonate, is secreted by the base of the coral (Figure 10–2). The reef-forming corals live in colonies that construct massive apartment house complexes to house the individual polyps.

Microscopic yellow algae known as zooxanthellae live within the outer flesh of the coral polyps. Individual algae cells are about the size of human red blood corpuscles. Corals and zooxanthellae live in a symbiotic relationship in which both benefit and neither is harmed. The corals provide a home for the zooanthellae, along with a steady supply of nutrients in the form of waste material, and carbon dioxide necessary for the plants' survival. The zooxanthellae assist the coral by removing waste material and absorbing excess carbon dioxide. The removal of carbon dioxide allows the coral to secrete calcium carbonate, which forms the limestone corallite. It has been suggested that without the zooxanthellae, the corals would not be able to secrete sufficient calcium carbonate to form the massive reefs. The zooxanthellae are plants and need light for photosyn-

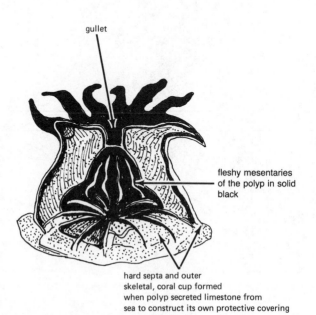

gullet

fleshy mesentaries
of the polyp in solid
black

hard septa and outer
skeletal, coral cup formed
when polyp secreted limestone from
sea to construct its own protective covering

FIGURE 10–2.
Cross-section of a coral polyp,
showing corallite, septa, and
tentacles.
Drawing by C. W. L.

thesis. Therefore, the presence of zooxanthellae in the coral polyps limits the reef-forming corals to shallow, well-illuminated waters of the tropics.

Some scientists have suggested that the zooxanthellae are a source of food for the corals. Although the idea of miniature built-in-farms might seem attractive for the corals, it is unlikely. When living corals are placed in a dark room, the zooxanthellae are ejected, but not eaten by the polyps. Other marine animals that secrete large shells, such as the giant clam, *Tridacna,* also contain zooxanthellae in their flesh (Figure 10–8). Unlike other clams, *Tridacna* faces upward with the shells open so that the zooxanthellae can soak in the beneficial rays of the sun.

The inside of the coral cup is divided by fleshy partitions that extend from the wall toward the center. The partitions increase the surface area on the inside of the cup, which allows the coral to digest food more rapidly. The fleshy partitions or mesenteries are supported by thin limestone walls known as septa, which project upward from the corallite. In colonial corals, the septa spill over the wall of the corallite and become fused with septa from the adjacent corallite. The area of fused septa is called the edge zone, and the colonial corals are known as edge-zone corals.

Septa projecting upward from the corallite are arranged like spokes in a wheel. The number of septa corresponds to the number of folds or partitions in the wall of the polyp. In the stony corals and sea anemones, the septa are arranged in sets of six. New septa are always added between the existing septa so that the total number of septa is an even multiple of six. Therefore, the stony corals and sea anemones are included in the hexacorals. The soft corals, including the sea fans and sea whips, have eight sets of septa and are known as octocorals.

Stony corals. The hard or stony corals that form the framework of the tropical reefs belong to the order Scleractinia, from the Greek word meaning hard. The stony corals have assumed a wide variety of different shapes that reflect their ecologic roles on the reefs. The six major shape categories include (1) branching corals, (2) brain corals, (3) platy corals, (4) encrusting corals, (5) mushroom corals, and (6) star corals.

The branching corals are the most abundant reef-forming corals in both the Atlantic and Indo-Pacific regions. Most of the branching corals belong to the genus *Acropora,* which live along the seaward edge of the reefs (Figure 10–1). There are more than 200 living species of *Acropora,* which represent about 40 percent of the stony corals. In the Pacific, a wide variety of different species of *Acropora* take on many different shapes, from short, stubby corals with fan-shaped branches to long, thin, delicate forms with complex corallites. In the Atlantic, only two species of *Acropora,* the staghorn coral *Acropora palmata* and the elkhorn coral *Acropora cervicornis,* are found on the reefs.

Acropora are considered the major reef-forming corals in both the Atlantic and the Pacific Oceans (Figure 10–1).

The brain or meandrine corals look like large, rounded boulders with furrows covering their surface (Figure 10–1). The individual polyps are lined up along the furrows, giving the coral the appearance of an animal brain. The polyps in the colony reproduce by budding so that the new infant polyps lie adjacent to the parents in long rows. The new polyps are not completely separated from their parents, and form a continuous line along the surface of the coral.

Although the mouths are separate, the bodies of the polyps are fused together. The tentacles of the brain corals are arranged in rows along each side of the grooves, and move in unison to bring food to the mouths. The genus *Diploria* is the most common brain coral in the Atlantic. Several different genera, including *Favia* and *Platygyra,* are abundant in the Indo-Pacific reefs.

The platy corals have polyps that are evenly distributed on thin sheets of calcium carbonate. The sheets are often folded to yield delicate patterns (Figure 10–1). Although the platy corals appear to be flowing with the waves, the sheets are rigid and easily broken off.

Encrusting corals live on thin sheets of calcium carbonate that are draped over other living masses of coral. The layers of encrusting corals often kill the polyps in the host coral. The fire coral, *Millepora,* also comes in mustard yellow sheets that cover areas on the surface of larger brain corals. The encrusting fire coral can inflict a painful wound.

Mushroom corals live in tide pools along the reef and in deeper water along the reef front. *Fungia,* the common mushroom coral on the Indo-Pacific reefs, is a solitary coral that lives between the larger coral heads (Figure 10–1). The septa on the mushroom coral radiate out from the center like the folds on the underside of a mushroom. The mushroom corals reproduce by budding small mushrooms along the edge. Instead of remaining attached, such as the buds in the brain coral do, the new mushroom corals drop off, and the parent continues to produce new offspring. The mushroom corals are also able to use their miniscule tentacles to move slowly about on the bottom, or to right themselves if they are flipped over.

The star corals are massive corals that are shaped like the brain corals, but the individual polyps are not connected in a row (Figure 10–29). In the star coral, each polyp has its own opening, with the star shape given by the septa converging inward toward the center of the corallite. The star coral *Montastrea annularis* forms large, knobby masses on the Atlantic reefs.

Soft corals. The soft corals have tiny polyps arranged along flexible arms, which wave back and forth with the waves and currents. Instead of having a solid corallite composed of limestone, the soft corals have intricate limestone needles aligned within flexible arms or branches. Although the soft corals are flexible when they are submerged in water, they harden when they are dried.

The soft corals come in several different shapes that are well described by the popular names: sea fans, sea whips, sea rods, organ pipe coral, umbrella coral, and many others. The sea fans and sea rods are the most familiar of the corals and are often called gorgonians (Figure 10–3). Both the sea fans and sea whips are attached to the bottom by a broad stem. The sea whip has several slender branches with thin featherlike arms and looks like a group of ostrich plumes. The sea fan, on the other hand, is arranged in a single plane with small openings for the polyps.

The soft corals are very colorful. They come in hues of yellow, orange, purple, red, violet, and lavender. When they are photographed with an underwater flash, the colors are vivid. If you soak the coral in bleach to remove the smell, the color will also fade. It is best to enjoy the soft corals in the reef, and to leave them for others to enjoy.

FIGURE 10–3.
The soft corals or gorgonians include sea fans and sea rods.
Drawing by C. W. L.

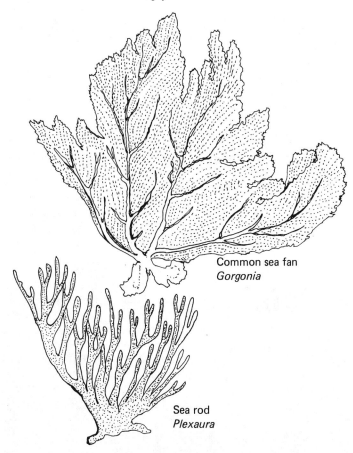

Common sea fan
Gorgonia

Sea rod
Plexaura

Coralline algae

Although the corals are obvious on the reefs because of their beautiful forms and intricate shapes, algae are more important to the basic economy of the reefs. It has been said that the corals are the bricks of the reef, but algae are the mortar that binds the coral together. Calcareous algae are minute plants with calcium carbonate within and between the cell walls and around the plant tissues. Calcareous algae are represented on the reefs by two major groups, *Chlorophyta* or green algae, and *Rhodophyta* or red algae.

Red algae on the reef front contain several genera, which are usually lumped together as Lithothamnion algae. The Lithothamnion algal group is identified by the vertical arrangement of cells separated by calcareous walls and connected with the outside by pores through the upper surface. Most coralline algae thrive near the water surface in high light intensity and warm water. The algae require clear water, and they can tolerate abnormally high salinity better than low salinity.

On existing reefs, the coralline algae are found on the outer reef slopes, the broad reef flat, and in the lagoon. Large colonies of branching red algae are abundant from the surface down to about 20 meters on the seaward slope. The calcareous algae reach their greatest abundance in the intertidal zone, forming an algal ridge. The algal ridge is an elevated strip of algae about 10 to 15 meters wide, which is kept moist by spray. Commonly, calcareous algae are the only organisms found on top of the ridge, but corals, mollusks, and echinoderms are also present in some of the protected pools and holes along the ridge.

The green algae, *Halimeda,* grows in clusters or patches on the reef flat and in the lagoon (Figure 10–4). Luxuriant *Halimeda* meadows have been described from the bottom of lagoons. The green algae break down into small needles of limestone, which form fine sand and mud in the lagoons. An active reef on an atoll could be considered analogous to a pail of water, with the reef and slope sediments forming the pail and the lagoonal sediments being the water within the pail. In total volume, the lagoonal sediments are usually much larger than the combined reef and slope sediments.

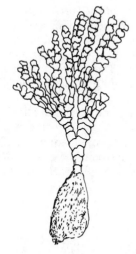

FIGURE 10–4.
The green algae, *Halimeda,* is abundant on the reef flat (length: 10 cm).
Drawing by C. W. L.

Reef-Dwelling Organisms

The reef-dwelling organisms include sponges, snails, clams, sea urchins, and sea stars, which live between the coral heads and within the sand ground up by the waves. In the lagoons behind the reefs, calcareous green algae grow profusely and supply fine needles of limestone to the sediment. Foraminifera are small, single-celled animals with delicate calcium carbonate shells, which live throughout the reef, but are especially abundant in the lagoon. Fish swim among the coral heads, and birds nest on the coral islands. Sea turtles struggle ashore to dig pits and lay their eggs on the coral islands known as cays.

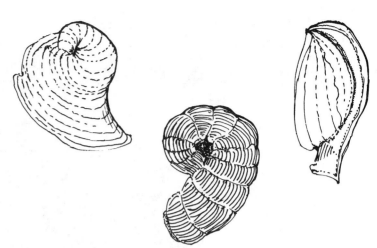

FIGURE 10–5.
Tests or shells of various shapes of foraminifera (magnified 10 times).
Drawing by C. W. L.

Foraminifera

Foraminifera are small, single-celled animals in the phylum Protozoa that are the dominant organisms on the shelf areas between the reefs. The forams have developed diverse shells and occupy many different habitats on the reefs (Figure 10–5). The shells form a large part of the bottom sediment in the interreef area and cover much of the ocean bottom above depths of about 5000 meters. At greater depths, the calcium carbonate shells are dissolved by carbon dioxide, which forms a dilute solution of hydrochloric acid. About 380 different species of forams have been identified on the Great Barrier Reef of Australia, and they provide a useful key to the different environments on the reef.

Sponges

Sponges are unusual aquatic animals that are abundant in the warm tropical waters around the reefs. There are more than 5000 species of sponges in the phylum Porifera that live in fresh water, and in the oceans from the tide pools to depths of 8500 meters.

The sponges are made up of individual protozoan cells that are gathered together to form sacklike bodies. In the sponges, the protozoan cells assume three different roles: (1) flattened external cells that are lined up to form the outer wall, (2) collar cells with whiplike tails, known as flagella, which line the internal canals, and (3) amoeboid cells that wander about in the jellylike mass between the outer walls and the internal cavity.

Sponges feed by pumping water in through pores in the walls and out through an opening in the top. The sponges are divided into three major types: simple tube sponges, tube sponges with radial canals, and complex sponges. The tube sponges have an upright tube lined with collar cells. The tube sponges with radial canals have fingerlike projections that contain the collar cells. The complex sponges have small cham-

FIGURE 10–6.
Cross-section through a large
barrel sponge, *Xestospongia*, from
the Caribbean (width: 70 cm).
Photo by W. T. Fox.

bers lined with collar cells between the outer wall and the inner chamber. The barrel sponge, *Xestospongia*, is a large, rotund sponge of the complex type that is common in the Caribbean (Figure 10–6). The barrel sponge grows up to 1.5 meters high and usually has a cone-shaped depression in the top. The outer wall is hard and knobby and resembles a large boulder.

The bodies of sponges are supported by organic fibers called spongin. In many varieties of sponges, small needles, called spicules, of calcium carbonate or silica are secreted within the spongin to give the sponges more rigid support. When the sponge dies, the spongin decays and the spicules are scattered about on the bottom. The common bath sponge consists entirely of spongin that has been dried out and had all the organic matter removed.

Clams and snails

Several different types of gastropods or snails are common on the Indo-Pacific and Caribbean reefs (Figure 10–7). The top shells represented by *Trochus* are abundant in the surf zone on the Great Barrier Reef of Australia. The ear shell *Teinotis* resembles common abalone with its distinctive ear shape and row of holes. The turbin shells *Turbo* are found on the algal ridge. The spider snail, *Lambis*, with ribs extending beyond the aperture lives on the sand flats and feeds on decaying vegetation. The large and beautiful tritons, *Charonia*, which inhabit the trumpet shell, are one of the few known predators on the crown-of-thorns starfish, *Acanthaster*. The cowrie shells that are abundant on the algal rim and the outer reef flat have a conspicuous well-rounded shape with a narrow-toothed aperture. The common yellow cowry, *Cypraea*, is about 3 centimeters long with a

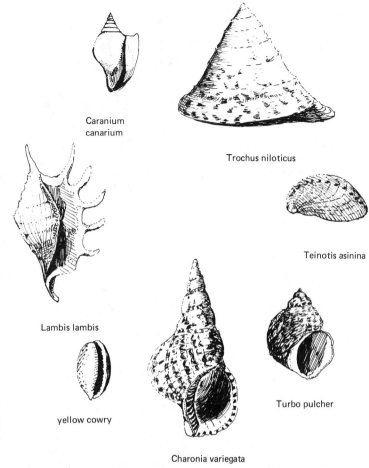

Caranium
canarium

Trochus niloticus

Teinotis asinina

Lambis lambis

Turbo pulcher

yellow cowry

Charonia variegata

FIGURE 10–7.
Common snails on the Great
Barrier Reef in Australia.
Clockwise from upper left: winged
shell, *Canarium*; top shell,
Trochus; ear shell, *Teinotis*; turbin
shell, *Turbo*; triton, *Charonia*;
cowry, *Cypraea*; and spider
shell, *Lambis*.
Drawing by C. W. L.

FIGURE 10–8.
Tridacna, a giant clam in a reef
surrounded by staghorn coral. It
can reach up to 4 feet in length
and weigh more than 400 pounds.
Drawing by C. W. L.

269

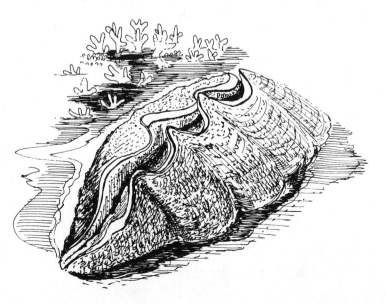

FIGURE 10–9.
A triton conch, *Charonia*, feeds on a crown-of-thorns starfish, *Acanthaster*, which in turn was feeding on coral polyps. A black sea urchin, *Diadema*, is behind. To the right, a *Stenopus* watches. Known for its bright coloration, this shrimp is commonly called the "barber-shop" shrimp.
Drawing by C. W. L.

dark brown color and covered by darker spots. Several other types of snails are also found in and around the coral reefs.

The pelecypods, or bivalves, are represented by *Tridacna*, the giant clam, which is up to a meter across. *Tridacna* has a purple mantle that appears to flow out of the shell (Figure 10–8). The color of the mantle is supplied by tiny plants that live within the flesh. Clams and oysters are also scattered throughout the reef front and on the reef flats.

Sea stars and sea urchins

Many species of sea stars can be found on the reefs, but the crown-of-thorns, *Acanthaster*, is the most spectacular (Figure 10–9). The crown-of-thorns is a spiral form that reaches more than half a meter in diameter. It feeds on coral polyps and has caused extensive damage to parts of the Great Barrier Reef. In 1966, it almost wiped out the coral on Green Island Reef. The explosion of *Acanthaster* has been blamed on people who collected triton snails, but now it appears that the crown-of-thorns has a natural cycle of explosive populations followed by more normal distributions.

The sea urchin, *Diadema*, is black with long spines and lives in groups on the reef flats (Figure 10–9). The sharp spines on *Diadema* can inflict a painful wound on an unwary swimmer. The Ophiuroidea, or brittle stars, are abundant on the reef surface, as well as in deeper water. The sea cucumbers, Holothuroidea, are common on the sand flats where they ingest the bottom sediment in search of food.

Crabs and shrimp

The phylum Arthropoda is represented by the crabs, shrimp, crayfish, barnacles, and sand hoppers. Many of the organisms, which were discussed in Chapter 9 are also present on the reef. The barnacles that covered broad areas of the rocky shore do not live on the coral, but are found on boulders in the algal rim and on the cemented calcareous sand that forms beach rock. Several gaudily colored varieties of crayfish live among the reef corals. The elusive hermit crabs scamper about the reef with snail shells upon their backs. Small crabs live among the delicate branching corals, and within the dome of the brain corals. Several different species of shrimp scamper about on the reef flats (Figure 10–9).

Reef fish   The coral reefs provide a protected habitat for many schools of fish, and other fish invade the reefs, looking for a meal. It is not possible to discuss here all the different species of fish that are encountered on a reef, but a few will be discussed that are potential hazards, and others that are closely interwined with the ecology of the reef.

The shark and barracuda are the largest and potentially most dangerous fish that swim through the reef (Figure 10–10). Sharks are unpredictable, and any shark, even the gentle nurse shark, can attack a swimmer. The tiger shark, with a broad, blunt snout, is the largest and most dangerous shark. Other sharks, including the hammerhead, bull, lemon, nurse, and sand sharks, swim onto the reefs to feed on the schools of fish. To prevent attracting sharks, it is advisable to boat all fish immediately and not to swim when you are bleeding.

The barracuda are among the most impressive fish that swim among the coral heads (Figure 10–11). They are generally about half a meter long, but can grow up to a meter and look like sleek, gray submarines. Barracuda have the habit of slowly opening and closing their imposing jaws as they hover close to a swimmer. Although some barracuda attacks have been reported, they will generally cause no harm. It is a much greater risk to eat barracuda, which are quite tasty but often bring along a case of food poisoning.

FIGURE 10–10. The bull shark, *Carcharhinis,* is among the largest predators on the reef.
Drawing by C. W. L.

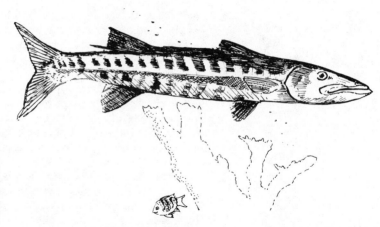

FIGURE 10–11.
The great barracuda, *Sphyaena*
barracuda, swims among the
coral heads (length: up to
1 meter).
Drawing by C. W. L.

Moray eels are fish that look like fearsome snakes poking
their jaws out of openings in the reefs (Figure 10–12). They have
long, narrow jaws with sharp teeth that can bite severely. Although the bite is not toxic, infection can follow from a moray
bite.

Rays often lie quietly in the shallow lagoon behind the reef
front, where they plow through the sediment in search of mollusks (Figure 10–13). Rays are strong swimmers, using their un-
dulating wings to fly through the water. Sting rays have spines
projecting upward from the base of their tails, which may contain poison. If you step on the spine, it can be quite painful, but
the rays are otherwise harmless.

A wide variety of different fish swim among the coral
heads, adding color and interest to the reef (Figure 10–14). The
jewfish and groupers are some of the largest fish on the reefs.
The jewfish, a giant sea bass, can grow to 2 meters and weigh
700 pounds. The grunts are an important source of food on the
reefs, and are noted for the grunting sounds they make underwater. Most of the grunts are black and white or blue and yellow, with stripes running lengthwise on their bodies.

FIGURE 10–12. Moray eels live in cracks or openings
in the reef front.
Drawing by C. W. L.

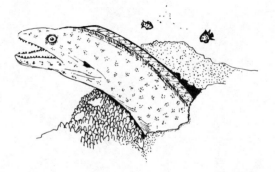

FIGURE 10–13.
Rays live in the quiet lagoon
behind the reef front—the
southern stingray.
Drawing by C. W. L.

FIGURE 10–14. Fish of the coral reef. A. Rock Beauty Angel
fish. B. Princess Parrotfish. C. Sargeant Major Damselfish. D.
Clown Wrasse. E. Spanish Hogfish. F. Porcupine fish.
Drawing by C. W. L.

The angelfish are the most colorful fish on the reefs, and their gaudy colors blend in quite well with sea fans and sea whips (Figure 10–14). The angelfish on the reefs grow considerably larger than the miniature variety we see in home aquaria. The damselfish is small but makes its presence known because it protects its territory where it guards purple or red egg clusters. The Sergeant Major, with its five black stripes over yellow background, will nip at larger fish, or even divers, when its territory is threatened. Wrasses are a diverse and colorful group of reef fish that undergo many changes in color and shape during their growth. Some of the small wrasses act as cleaner fish and pick the parasites off of larger fish, who patiently wait for the service. The larger wrasses, such as the hogfish with long, pig-like snouts, can reach up to a meter in length and are much sought after as food.

The parrot fish, with their stubby jaws, graze on algae that live on the corals. In the process, they nibble off bits and pieces of coral and grind it into fine sand. The parrot fish can eat a large volume of coral, keeping the growth of the reef in check and adding to the sand supply for the beaches. Although the common blue parrot fish is a blue-green color, other parrot fish are brightly colored with rainbow and striped patterns.

The triggerfish have small mouths but powerful jaws that enable them to eat invertebrates. Their eyes are set well back from their mouths, which prevents damage from long spines when they eat sea urchins. When the triggerfish hides in a crevice in the reef, the first dorsal fin is locked in place by the second fin so the fish cannot be dislodged.

The puffer, or porcupine, fish looks harmless but can be deadly poisonous. The puffers have the uncanny ability to blow up like balloons to about three times normal size, by drawing water into their bellies. Their tough skin is covered by spines that extend outward when they are inflated. Although some puffers are considered a delicacy in Japan, their internal organs contain a deadly poison and can be fatal if improperly cleaned.

If you are planning to snorkel or dive on a reef, the water-proof *Guide to Corals and Fishes* by Idaz and Jerry Greenberg is an excellent companion. The colored drawings of the fishes and corals make rapid identification easy. Although it is difficult to actually read the book while swimming underwater, it is handy to carry the guide while you swim and to pop up to the surface for a quick look at a picture of the fish or coral you have just seen underwater.

Indo-Pacific Coral Reefs

Coral reefs are found in tropical and subtropical climates in a broad belt across the Indian, Pacific, and Atlantic Oceans between about 30° north and 30° south latitude (Figure 10–15). Coral reefs can be divided into two broad groups: oceanic reefs that form atolls and fringing reefs around volcanic islands, and shelf reefs on islands and flanks of continents.

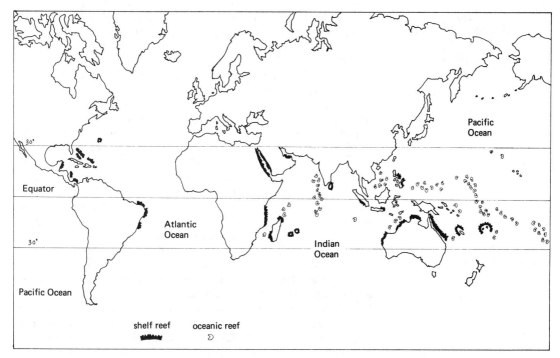

Pacific
Ocean

30°

Equator

Atlantic
Ocean

30°

Indian
Ocean

Pacific Ocean

shelf reef   oceanic reef

FIGURE 10–15. Coral reefs exist only in tropical and subtropical waters. Thus they are limited to appearing solely around the earth's midsection, extending 30 degrees north and south of the equator.
Drawing by C. W. L.

Fringing, barrier, and atoll reefs occur on the hundreds of volcanic islands scattered throughout the Indian and Pacific Oceans. Atolls are rings of coral that exist above many submerged volcanoes. The largest reefs in the Indo-Pacific belt occur in the Great Barrier Reef on the northeast coast of Australia. In the western Atlantic, the Caribbean shelf reefs extend from Florida to the Bahama platform and south to several small islands around the Caribbean Sea.

The distribution of reefs is limited by the biological requirements of the organisms living on the reefs. The reef-building corals require warm, relatively shallow, clear saline water. Surface currents that flow from east to west along the equator supply warm, clear water to the coral reefs along the western margins of the ocean. The coral larvae are also spread from island to island by the warm currents.

Large rivers inhibit the growth of coral reefs by introducing fresh water and suspended sediment. Fresh water from the rivers lowers the salinity below the tolerance of the corals. Reduction of salinity below about 80 percent of normal salinity (35 parts per thousand of salt in the water) will affect the life of the corals. Sediment carried to the sea by large rivers results in turbidity, which cuts down on the amount of light that penetrates into the water. The excess sediment also clogs the pores of small corals and eventually causes their demise.

Oceanic reefs and atolls

Atolls are more or less continuous oval, ringlike, or oblong reefs that surround a deeper lagoon. Coral atolls are quite extensive in the Indian and Pacific Oceans, where more than 300 atolls have grown on extinct volcanoes. Broad areas of deep ocean within the western Pacific also contain basaltic volcanoes. The basaltic volcanoes that rose from the ocean floor have slowly subsided beneath the surface of the ocean. As the Pacific Plate moved to the west away from the spreading center in the east Pacific, the Pacific Plate cooled and settled, lowering the level of the ocean floor (see Chapter One). The massive volume and weight of the individual volcanoes caused them to subside within a circular moat or depression in the ocean floor.

In the Indian Ocean, fringing reefs occur along the east coast of Africa, and around several of the granitic islands, including Madagascar and the Seychelles. The Laccadive, Maldive, and Chagos Archipelagoes are clustered along a ridge south of India. The ridge appears to be a scar left on the ocean floor from when the continent of India moved northward and collided with Asia. Suvadiva Atoll in the Maldives is one of the largest atolls in the world, having a length of 70 kilometers and a width of 58 kilometers.

Spurs and grooves. Along the edge of the atolls, nearly straight grooves are present at right angles to the reef front. On Bikini Atoll, the grooves are 1.5 to 3 meters wide and about 8 meters deep at their inner ends (Figure 10–16). The grooves are separated by spurs of living coral or calcareous algae. Some of the grooves extend into the algal ridge as surge channels, reaching 100 meters into the reef. Algae often grow over the surge channels, forming tunnels. Before the roof is complete, blowholes often develop along the surge channel. Encrusting algae build mounds around the blowholes, forming geyserlike algal craters. Where the surge channels are completely roofed over, a honeycomb structure is formed beneath the algal pavement.

The grooves play an important role in the development of the reef and act as an efficient breakwater. Large breakers are generated by the trade winds on the windward side of the reef. When one looks seaward, a breaker appears to be crashing toward the shore with full force. Two seconds later, the wave has disappeared as a noisy surge with the groove. It has been estimated that about 95 percent of the wave energy striking the reef is dissipated within the grooves. The total wave power on the windward side of Bikini Atoll averages 500,000 horsepower per day, one-fourth the power capacity at Hoover Dam.

Coral zonation on atolls. Bikini Atoll is a typical example of a Pacific oceanic reef. It is located in the Marshall Islands

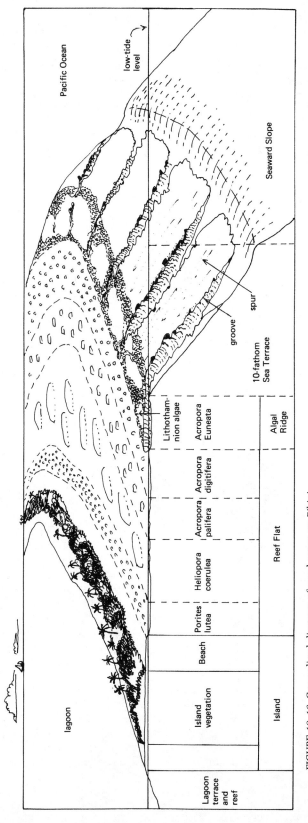

FIGURE 10–16. Generalized diagram of coral zones on Bikini Atoll with seaward grooves and spurs. *Drawing by C. W. L.*

about 4,200 kilometers (2,500 miles) southwest of Honolulu. Bikini is about 40 kilometers long and 25 kilometers wide. The lagoon covers an area of about 700 square kilometers and has an average depth of 45 meters. Most of the islands are 3 to 4 meters above sea level, but parts of Bikini Island are 5 to 6 meters high.

The windward reef at Bikini Atoll has been divided into a set of bands or zones parallel to the beach (Figure 10–16). The *Acropora cuneata* zone includes the algal ridge, which is typical of the windward margins of many Pacific reefs. The *Acropora digitifera* zone contains a number of microatolls with the mushroom coral, *Fungia. Acropora digitifera,* the distinctive coral of this zone, has slender, finger-sized branches and bright, fluorescent-green or grayish polyps.

The *Acropora palifera* zone includes the reef flat, a band 15 to 90 meters wide. The *Heliopora coerula* zone is submerged by about a meter of water at low tide and varies in width from 15 to 150 meters. The microatolls form almost a continuous surface over which one can walk at low tide. The *Porites lutea* zone lies adjacent to the beach. Yellow and purplish-brown variants of *Porites lutea* are the only living corals (Figure 10–16). The beach forms a narrow rim of coral and foraminiferal sand around the small islands. The broken coral fragments are ground up by the waves to form the white coral sand beaches.

Subsidence theory of atolls—Darwin. In 1832, Charles Darwin explored and mapped one of the atolls in the Indian Ocean on the voyage of the *Beagle.* He made several profiles across the Keeling or Cocos Atoll in the eastern Indian Ocean south of Sumatra. Based on his studies of the coral reefs in the Indian Ocean, and from the descriptions of reefs by other travelers, Darwin published a book, *The Structure and Distribution of Coral Reefs,* in 1842. Although Darwin is better known for his work on the evolution of species, his book on coral reefs is a scientific classic and his theory on the origin of atolls still holds today. Minor modifications have been made to his theory, based on later evidence that was not available to Darwin.

In his book on coral reefs, Darwin included a map showing the worldwide distribution of fringing reefs, barrier reefs, and atolls. Fringing reefs are located along the shore of large islands or along the edge of a continent. Barrier reefs are separated some distance from the coast by a shallow lagoon. An atoll is a ring of small coral islands and submerged reefs that surround a lagoon. Darwin used subsidence to account for the evolution of a fringing reef into a barrier reef and eventually into an atoll with a broad lagoon (Figure 10–17). When volcanic activity ceased, the shallow shoreline surrounding the volcano became populated with corals. Coral larvae floating in the ocean current settled along the shore, where they found a suitable substrate. As the corals grew, they formed a fringing reef around the volcano. As the volcano settled under its own weight, the

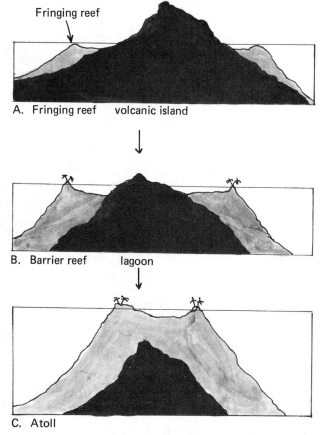

A. Fringing reef   volcanic island

B. Barrier reef   lagoon

C. Atoll

FIGURE 10–17. Darwin's theory of progressive subsidence: There is a progression from *A* to *C* as the sheer weight of volcanic rock sinks the island beneath the water surface and causes the coral reef growth to accumulate upward.
Drawing by C. W. L.

reef continued to grow upward, forming a barrier reef. When the central island eventually sinks beneath the level of the lagoon, the barrier reef becomes an atoll. Darwin supported his theory by citing maps of volcanic islands with fringing, barrier, and atoll reefs in various stages of evolution. He surmised that the atolls would be capped by a thick layer of shallow water limestone representing countless generations of reefs.

Glacial control of atolls—Daly. After the turn of the century, a prominent geologist, Reginald A. Daly, disagreed with Darwin and proposed the glacial theory for the control of coral atolls. According to Daly's theory, coral atolls grew around the edges of volcanoes that had been planed off during the ice age (Figure 10–18).

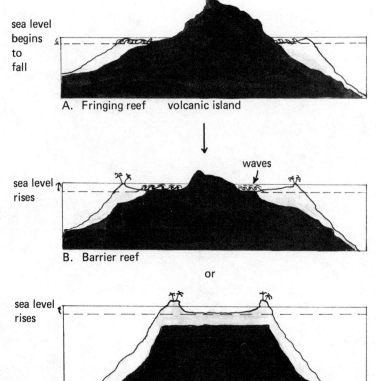

sea level
begins
to
fall

A. Fringing reef   volcanic island

waves

sea level
rises

B. Barrier reef

or

sea level
rises

C. Atoll

FIGURE 10–18.
Daly's theory of glacial control of
atolls: A. During the ice age when
the sea level sank, wave action cut
into the island. B. Subsequent sea
level rise caused either a barrier
reef to remain, or C. an atoll to
develop as favorable conditions for
reef colonization resulted.
Drawing by C. W. L.

Daly observed that almost all the lagoons within the atolls
had a nearly uniform depth of between 20 and 25 meters (60 to
75 feet). When large portions of North America and Europe
were covered with glacial ice about 25,000 year ago, sea level
was substantially lower. At that time, the tops of many of the
Pacific volcanoes were planed flat by wave action. Coral reefs
become established around the edges of the oval platforms.
When the glaciers melted and sea level rose, about 10,000 years
ago, the corals grew upward to keep pace with sea level. Therefore, the atolls were formed, with coral reefs and islands surrounding lagoons with remarkably uniform depths.

Drilling to test atoll theories. The obvious test to determine if the atoll was the product of subsidence, or was controlled by changes in sea level due to glaciation was to drill a
hole in an atoll. Following World War II, the United States
drilled several holes on Bikini and Eniwetok Atolls in preparation for atomic bomb tests. On Eniwetok Atoll, two deep holes
were drilled through limestone into basalt at 1268 and 1405

meters. Shallow-water corals were encountered near the bottom of both holes.

The oldest fossils at the base of the deepest hole are of Eocene age, about 60 million years old. Sea level remained constant for a long period of time, while the top of the volcano was eroded almost flat. Slowly, the volcano started to subside and the coral growth kept pace with the subsidence.

There are three leached zones within the 1200 meters of limestone, indicating that the volcano stopped subsiding and was uplifted three times during the formation of the atoll (Figure 10–19). Leaching takes place when fresh water charged with carbon dioxide dissolves some of the limestone. The deep drill holes indicated at least 1200 meters (4000 feet) of subsidence, which gave substantial support to Darwin's theory of coral reef formation.

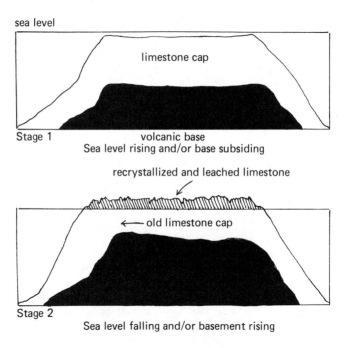

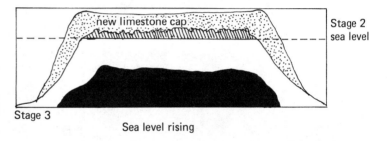

FIGURE 10–19.
Leached zones in the reef limestones on Eniwetok Atoll indicate that the volcano stopped subsiding and was uplifted during formation of the atoll.
Drawing by C. W. L.

The Great Barrier Reef of northeastern Australia is the largest and most prolific reef area in the world. The reefs occur on Queensland Shelf between 10° and 24° south latitude, a distance of 1600 kilometers (1000 miles) (Figure 10–20). At Cape Melville, near the north end, the shelf is 24 kilometers (15 miles) wide, and it broadens out to 290 kilometers (180 miles) at Cape Townshend. Wall reefs grow along the seaward margin of the shelf, where nutrient-rich, warm water floods the coral reefs. Platform reefs grow on the impoverished inner shelf behind the wall reefs.

Physical conditions on the Great Barrier Reef. The Great Barrier Reef is ideally situated for the productive growth of coral polyps. The basic requirements for a reef are (1) shallow water where light can penetrate to the bottom, (2) warm temperature, (3) sufficient nutrients for food supply, (4) constant aeration for the removal of waste material, and (5) an adequate source of carbonate ions in the water for building the external skeleton of calcium carbonate. If one of these factors falls below the minimum requirement, the reef will not prosper. It is sometimes difficult to satisfy all the requirements simultaneously, because warm water is usually deficient in carbonate ions. For example, a carbonated beverage is cooled to retain the carbon dioxide. When the drink warms up, the gas escapes and the drink becomes flat.

The Queensland Shelf satisfies the requirement for shallow water. In the northern region, from 10° to 16° south latitude, the shelf has an average depth of 30 meters (96 feet) and coral reefs are scattered across the shelf. In the central region, from 16° to 21° south latitude, the inner part of the shelf is 35 to 55 meters (120 to 180 feet) deep and the reefs are concentrated along the seaward margin of the shelf. In the southern region, from 21° to 24° south latitude, the shelf has an average depth of more than 55 meters (180 feet) and the reefs are clustered along the edge of the shelf.

Warm water for the reefs is supplied by the fast-moving South Equatorial current. The current flows across the Pacific south of the equator and is heated along the way. When the current reaches the Coral Sea to the east of Australia, it splits into two parts, the East Australia current that heads to the south along the coast of Australia, and a northern current that flows into the Solomon Sea. The Equatorial Current that bathes the Great Barrier Reef has an average temperature of 27 degrees Centigrade (80.6 degrees Fahrenheit).

Nutrients for coral growth are supplied by upwelling. When the south equatorial current splits in two, the divergence on the surface produces an upwelling of deep nutrient-rich water in the Coral Sea. The Coral Sea has an average temperature of 21.3 degrees Centigrade (70.3 degrees Fahrenheit). The nutrient-rich water provides an increase in phytoplankton,

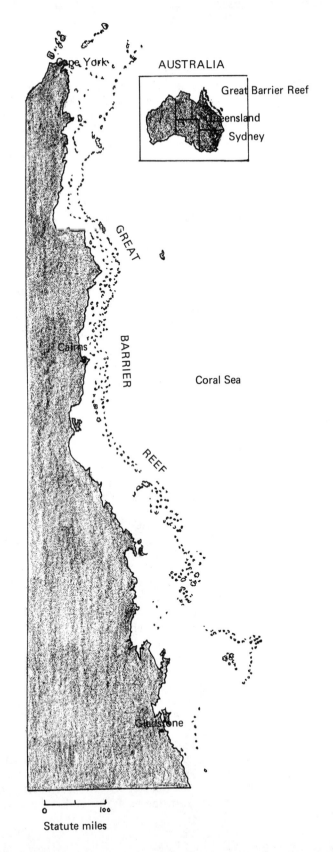

FIGURE 10–20.
The Great Barrier Reef along the
northeast coast of Australia.
Drawing by C. W. L.

which in turn results in an increase in the zooplankton. The outer reefs benefit from the food supply, while the reefs on the inner shelf receive a depleted food supply.

Waves and tides provide the necessary aeration for the Great Barrier Reef. The southeast trade winds blow from May to November, providing a steady surf along the outer reefs. In the summer, from December to April, the summer monsoon blows from north of the Equatorial trough. The tides along the Queensland Shelf have an average range of 3 to 4 meters (10 to 13 feet). In the central section of the reef at Broad Sound, two tide systems converge and the tidal range reaches 10.4 meters (34 feet). The tides generate strong tidal currents that ebb and flood between the reefs along the edge of the shelf. The steady waves and strong tidal currents stimulate coral growth and remove waste material.

The carbonate ions required for growth of the limestone foundation of the reefs are supplied by the upwelling water in the Coral Sea. The carbon dioxide–rich water from the Coral Sea is mixed with the warm water of the South Equatorial Current to provide just the right combination of temperature, nutrients, and carbon dioxide for luxurient reef growth.

Wall reefs on the outer shelf. Along the outer edge of the Queensland Shelf where the bottom drops off rapidly to oceanic depths, wall reefs develop parallel to the edge of the shelf. The wall reefs face into the warm South Equatorial current and are washed by the waves from the southeast trade winds. The spur and groove system is well established along the seaward edge of the wall reefs. Many different species of the branching coral *Acropora* grow along the grooves.

The wall reefs go through a series of stages, which can be seen along the Great Barrier Reef (Figure 10–21). Where strong tidal currents sweep around the ends of the wall reefs, the reef growth forms a spitlike shape at each end and becomes a cuspate reef. The cuspate reef will develop into an open ring reef and finally a closed ring reef (Figure 10–21). Where waves wash over the top of the wall reef, a prong reef develops with narrow buttresses extending behind the main wall reef. If the prong reef becomes enclosed by cusps, an open mesh reef forms, followed by a closed mesh reef (Figure 10–21).

Platform reefs on the inner shelf. Platform reefs develop on the inner shelf behind the wall reefs, where they are cut off from the open ocean surf. The platform reefs have a circular or oval shape where the wave and current action does not favor one side. The lagoon and reef flat have a saucer-shaped profile with a lifeless lagoon in the center.

The zonation on the platform reefs is well displayed on Heron Reef (Figure 10–22). A rim of calcareous algae is slightly raised above the rest of the reef flat. The reef flat extends inward from the algal rim and is divided into three distinct zones. The

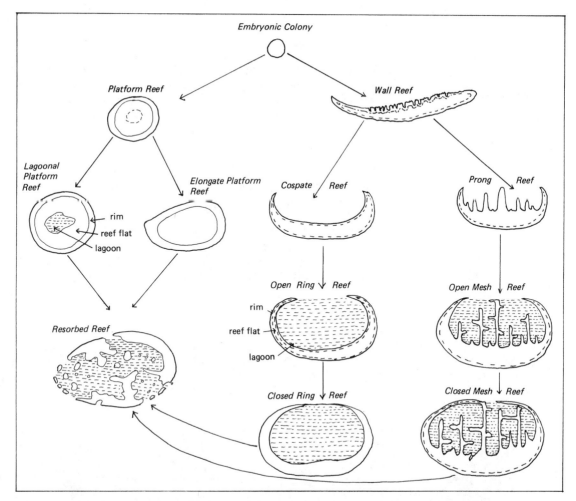

FIGURE 10–21. Evolution of two major reef types: wall and
platform of the Great Barrier Reef Province.

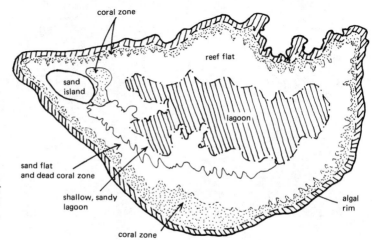

FIGURE 10–22.
Heron Reef, an example of reef
zonation on a platform reef on
the Great Barrier Reef Province.

zone of living corals and coral pools thrives in the protected area behind the rim (Figure 10–22). Dense populations of branching corals, including *Acropora, Pocillopora,* and *Seriatopora,* are nourished by water that surges up the grooves and tunnels beneath the algal rim (Figure 10–23). The black long-spined echinoid, *Diadema,* is also abundant in the coral pools.

The sand flat and dead coral zone is marked by an abrupt change from branching corals to brain corals and sand. Broken fragments of branching corals, boulders, and large slabs of beach rock are scattered across the sand flat. Small sand islands, known as sand cays, often form at the windward and leeward ends of the reef flats (Figure 10–22). Most of the islands are invaded by bird populations and are often the nesting grounds for the Silver Gull. Several different species of terns, including Noddy, Bridled, Crested, Roseate, and Sooty Terns, inhabit the cays. Sea turtles come ashore onto the beaches to nest and lay eggs (Figure 10–24).

FIGURE 10–23. Several species of branching and brain corals are found on the reef flats on the outer reefs.
Courtesy of Australian Tourist Commission.

FIGURE 10–24.
Green turtles of the Great Barrier Reef in Australia have been so hunted that their numbers have been precariously reduced.
Drawing by C. W. L.

Atlantic Reefs

The Caribbean Sea is a westward extension of the Atlantic Ocean, bounded by Cuba and Hispaniola on the north; Mexico and Honduras on the west; Panama, Colombia, and Venezuela on the south; and the Lesser Antilles on the east. Typical West Indies coral reefs are found in the Caribbean and along the neighboring coasts of south Florida and the Bahama Islands. Some reef corals occur on shallow knobs in the Gulf of Mexico, but mud, silt, and sand from the Mississippi River inhibit coral reef growth along the Gulf Coast. The Orinoco and Amazon Rivers also limit reef growth south of 5° north latitude on the western side of the Atlantic.

Caribbean corals

The coral reefs in the Caribbean are limited to about twenty-five genera and fifty species of reef-forming corals. The most abundant genera in almost all the reefs include six stony corals— *Acropora, Montastrea, Porites, Diploria, Siderasterea,* and *Agaricia*—and one hydrozoan—*Millepora*— which account for more than 90 percent of the total coral biomass. Abundant soft corals, including sea fans and sea whips, are found among the hard corals in the Caribbean reefs. It is much easier to study corals in the West Indies than in the Indo-Pacific region because the number of species is limited, and the same few species are present on almost every reef (Figure 10–25).

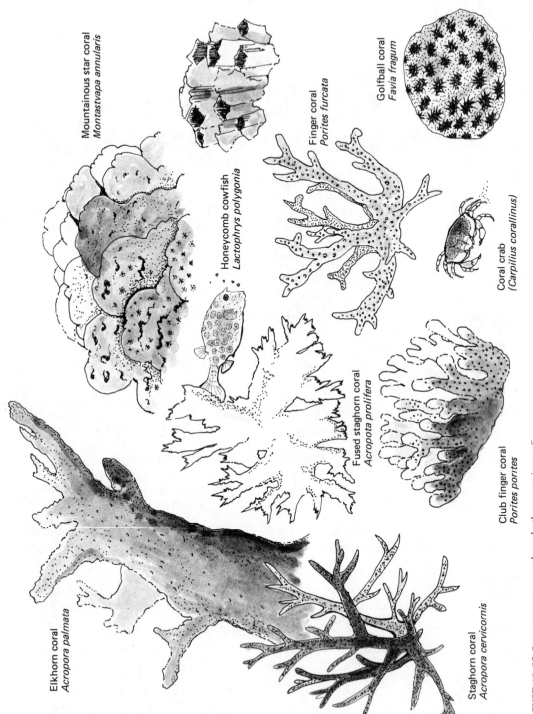

Mountainous star coral
Montastvapa annularis

Finger coral
Porites furcata

Golfball coral
Favia fragum

Honeycomb cowfish
Lactophrys polygonia

Coral crab
(*Carpilius corallinus*)

Fused staghorn coral
Acropota prolifera

Club finger coral
Porites porites

Elkhorn coral
Acropora palmata

Staghorn coral
Acropora cervicornis

FIGURE 10–25. Some common corals and other organisms of the Caribbean Reefs of the Atlantic. *Drawing by C. W. L.*

The moosehorn or elkhorn coral, *Acropora palmata,* is the most common branching coral in the Caribbean. The elkhorn coral has small corallites that are spread over broad, flat branches (Figure 10–26). It is a light brown or tan color and thrives in the well-oxygenated surf zone facing the incoming waves. Several growth forms of elkhorn coral are found in the Caribbean. On the reef front facing the waves, the branches are lined up, heading into the approaching waves (Figure 10–26). In deeper water, the branches grow upward in a random fashion, while in very shallow water, the coral has more massive form with thicker trunk and branches.

The staghorn or deerhorn coral, *Acropora cervicornis,* has thinner branches that group together, forming a thicket of coral (Figure 10–25). The staghorn coral is a more fragile coral than the elkhorn and is frequently broken off by heavy surf. *Acropora cervicornis* often grows in more protected areas behind the reef front and in deeper water where it can not be battered by the waves.

FIGURE 10–26. Elkhorn coral, *Acropora palmata,* grows with branches facing the waves (depth: 3.5 meters).
Photo by K. Ruetzler, Smithsonian Institution.

FIGURE 10–27. Mountainous star coral, *Montastrea annularis,*
with sea fans and sea whips in the Spur and Groove Zone,
Carrie Bow Cay, Belize.
Photo by K. Ruetzler, Smithsonian Institution.

The club finger coral, *Porites,* has short, stubby branches
with tiny corallites flush with the surface of the branch (Figure
10–25). The club finger coral is a fragile coral that often grows
with the staghorn coral in protected areas of the reef.

The mountainous star coral, *Montastrea annularis,* forms
large, knobby masses on the reef front and extends down to
depths of about 25 meters (Figure 10–27). On the four atolls in
the southwestern Caribbean off the coast of Belize, the star coral
makes up about 60 percent of the coral biomass. However, in
the northern Caribbean reefs off Florida and the Bahama Islands, elkhorn coral, *Acropora palmata,* is the dominant coral.
On the deep forereef at Discovery Bay on the north coast of
Jamaica, *Montastrea cavernosa* forms a reef at a depth between 55 and 70 meters, one of the deepest living reefs in the
Caribbean.

The brain coral, *Diploria,* has three species that are abundant on the reefs. The common brain coral, *Diploria strigosa,*
has its polyps arranged in rows along the surface of the coral
(Figure 10–28). The septa from adjoining rows are in contact,
giving the appearance of worms crawling about on the top of

FIGURE 10–28. Brain corals, *Diploria strigosa* and *D. clivosa*, at a depth of 8 meters on Carrie Bow Cay, Belize. *Photo by K. Ruetzler, Smithsonian Institution.*

the coral. The encrusting brain coral, *Diploria clivosa,* forms a sheet or layer on top of other corals or blocks of broken coral debris. The third species of brain coral, *Diploria asteroides,* has a line down the middle of the septal ridge, which forms a distinctive groove. The brain coral favors a protected habitat with the star and staghorn corals.

The starlet coral, *Siderastrea,* and the leaf or pineapple coral, *Agaricia,* are common small corals on the reef flat (Figure 10–25). At low tides, some of the coral heads on the reef flat protrude above the water surface.

The hydrozoan, *Millepora,* which is known as the stinging or fire coral, grows on the shallow reef flats. The most common growth form of *Millepora* is thin, vertical sheets with curving walls (Figure 10–25). However, it can also form a yellow encrusting layer on species of brain coral. The sting from *Millepora* can vary from a slight tingling sensation to a painful wound.

Caribbean reefs

291

Coral reefs are found in shallow water surrounding many of the Caribbean Islands. San Salvadore Island in the Bahamas, where

Columbus landed on October 12, 1492, is noted for its beautiful reefs. The large islands, Jamaica and Puerto Rico, also have reefs along their shore. The Antilles form a chain of islands, including the leeward and windward islands. The U.S. Virgin Islands, Saint Thomas and Saint Croix, are in the leeward islands. The windward islands to the southeast include Saint Lucia and Barbados, which are renowned for their reefs. Aruba, Curaço, and Bonaire are located off the coast of Venezuela. To the northeast, Roatán off the coast of Honduras, and Rendezvous Cay adjacent to Belize, have well-preserved reefs.

Florida reefs. Coral reefs are present along the Florida Keys from Key Largo to the Dry Tortugas. The most accessible reefs with extensive corals are located at Key Largo Dry Rocks. The flourishing reefs occur near the edge of the shelf on the seaward side of the emergent keys. Sediment from the Florida Bay to the north of the Keys tends to inhibit coral growth. However, the islands on the Keys block the southward flow of sediment and encourage active reef growth.

Spur and groove structures are present along the seaward edge of the reefs. The spurs range from 6 to 30 meters wide and reach depths of about 30 meters. The flanks of the spurs are lined with elkhorn coral and the stinging coral, *Millepora,* is abundant on the tops of the spurs. Sections of the spurs were blasted to determine the method of their growth. On the Florida reefs, evidence indicates that the spurs have enlarged seaward by biologic growth and are not the product of erosion in the grooves. Return currents carry sand and broken coral fragments down the grooves, which prevent coral attachment. Heavy seas during hurricanes clean off the spurs and deposit the accumulated debris in the grooves.

Boats and snorkeling gear are available at the John Pennekemp Coral Reef at Key Largo, Florida. Many different types of coral, sponges, and fish typical of the Caribbean reefs can be seen at the Pennekemp Reef.

Buck Island Reef—Saint Croix. At Saint Croix in the U.S. Virgin Islands, the National Park Service maintains the only underwater park, Buck Island National Reef Monument. Although SCUBA diving is not allowed in the park, because the SCUBA tanks can damage the corals, it is the ideal place to snorkel. Arrow markers and signs on the ocean floor guide snorkelers along the Buck Island Reef nature trail. The major coral types including *Acropora, Montastrea,* and *Diploria* are clearly labeled with underwater signs. The fire coral, *Millepora,* dominates the shallowest portions of the reef top. Abundant schools of fish swim among the coral heads. Along the reef front, the swimmer is carried back and forth by the surging waves past the oriented branches of *Acropora palamata. Acropora cervicornis* and *Agaricia* are abundant in the back reef area. Boats with snorkeling equipment on board are available to Buck Island from Saint Croix. The warm, clear waters and beautiful reefs provide all the necessary ingredients for a wonderful day of reef watching.

Grand Bahama Banks. The Grand Bahama Banks east of Miami, Florida, contain more than 6000 square kilometers of shallow-water limestone environments. The coral reefs are located on the eastern edge, where nutrient-rich water floods up over the banks. Behind the reefs, extensive areas of lime sand and mud are formed. The carbonate environments on the Bahama Banks are of great interest to petroleum geologists because they provide a modern analogue for fossil limestone deposits that they encounter on the continent. The vast oil reservoirs of the Middle East were formed in deposits similar to the Bahama Banks.

The northern part of the Bahama Banks are covered with sheets of oolitic sand. Oolites are small, sand-size grains that have concentric layers of calcium carbonate deposited around a nucleus that is usually an aragonite needle. The oolites are formed in the surf zone on the edge of the bank and carried west by the currents to form large underwater dunes of oolites.

To the west of Andros Island, the banks are covered by lime and pellet muds. The muds are the product of the breakdown of the green algae, *Halimeda* (Figure 10–5) and the pellets are extruded by snails and other small organisms.

The Bahama Banks provide an unusual outdoor laboratory for the study of limestone in the process of formation. Although the average depth on the banks is between 10 and 20 meters, the limestones are several thousand meters thick. During the Pleistocene glaciation, the entire Bahama platform stood well above sea level and formed a desolate plain of lime mud, and sand. Now, it provides a thriving underwater garden of coral and algae.

Summary

Coral reefs are limited to the tropics between 30° north and 30° south latitude. The reefs are most abundant on the western sides of ocean basins, where warm currents have a temperature between 23 degrees and 25 degrees Centigrade. The depth of the reef-building corals is limited by the yellow algae zooxanthellae, which live in the outer flesh of the corals.

The framework of the reefs consists of stony corals and calcareous algae. Soft corals and sponges live attached to the bottom between the corals. Clams, snails, sea stars, sea urchins, crabs, and shrimp live on the corals and roam the bottom in the reefs. Brightly colored tropical fish swim among the coral heads.

In the Indo Pacific, oceanic reefs and atolls develop on volcanic islands, and shelf reefs on the margins of continents. The atolls are formed on subsiding volcanoes, which are surrounded by upward-growing reefs. The Great Barrier Reef on the coast of northeastern Australia is the largest reef tract in the world.

The Atlantic reefs in the Caribbean Sea extend from Florida to the northern coast of South America. The smaller number of stony coral species and increase in soft corals distinguish the Atlantic from the Indo-Pacific reefs.

Selected Readings

ADEY, W., W. GLADFELTER, J. OGDEN, AND R. DILL. *Field Guidebook to the Reefs and Reef Communities of St. Croix, Virgin Islands.* Miami, Fla.: University of Miami, 1977. An excellent field guide to the coral reefs at Buck Island National Monument.

DARWIN, C. *The Structure and Distribution of Coral Reefs.* Berkeley, Calif.: University of California Press, 1962. A paperback version reprinted from the original 1851 classic on the formation of coral reefs.

DAVIS, W. M. *The Coral Reef Problem.* New York: American Geographical Society, 1928. An interesting review of the different hypotheses for the origin of coral reefs.

GREENBERG, I. AND J. GREENBERG. *Waterproof Guide to Corals and Fishes of Florida, Bahamas and Caribbean.* Miami: Seahawk Press, 1977. A handy guide for snorkling among the reefs.

JACOBSON, M. K. AND D. FRANZ. *Wonders of Corals and Coral Reefs.* New York: Dood, Mead and Company, 1979. An excellent introduction to coral reefs for the general public.

JONES, O. A. AND R. ENDEAN. *Biology and Geology of Coral Reefs.* New York: Academic Press, 1973. A collection of chapters by scientists on different reef provinces.

LAPORTE, L. (ed.). *Reefs in Time and Space.* Tulsa, Okla.: Society of Economic Paleontologists and Mineralogists, 1974. A collection of original papers on living and fossil reefs.

MAXWELL, W. G. H. *Atlas of the Great Barrier Reef.* Amsterdam: Elsevier Publishing Co., 1968. An excellent description of the physical and biologic factors that influence the Great Barrier Reef.

MULTER, H. G. *Field Guide to Some Carbonate Rock Environments, Florida Keys and Western Bahamas.* Dubuque, Iowa: Kendall/Hunt Publishing Co., 1977. A complete guide and description of the Florida reefs.

POWER, A. *The Great Barrier Reef.* New South Wales, Australia: Paul Humlyn Pty. Ltd., 1969. A beautiful collection of color photographs of fish, corals, and other animals on the Great Barrier Reef.

RUETZLER, K. AND I. MACINTYRE, (eds.). *The Atlantic Barrier Reef Ecosystem at Carrie Bow Cay, Balize, 1.* Washington, D.C.: Smithsonian Institution, 1982. An up-to-date scientific account of the only true barrier reef in the Caribbean Sea.

STOKES, F. J. *Handguide to the Coral Reef Fishes of the Caribbean and Adjacent Tropical Waters Including Florida, Bermuda and the Bahamas.* New York: Lippincott and Crowell, 1980. A pocket guide to 460 species of fish found on coral reefs.

WIENS, H. *Atoll Environment and Ecology.* New Haven, Conn.: Yale University Press, 1962. A good review of atoll environments and formation.

Glossary

ADVECTION FOG. Coastal fog caused by horizontal movement of warm, moist air across cold water.

ALGAE. Marine or freshwater plants, including phytoplankton and seaweeds.

AMPHIDROMIC POINT. The center of an amphidromic tidal system; a nodal or no-tide point around which the crest of a standing wave rotates once each tidal period.

ANTINODE. That part of a standing wave where vertical motion is greatest and horizontal motion is least.

APHELION. The point on the earth's orbit most distant from the sun.

APOGEE. The point on the moon's orbit most distant from the earth.

ARAGONITE. A carbonate mineral similar to calcite $CaCO_3$ but with a different crystal structure.

ARTHROPODS. Animals with segmented external skeleton of chiton or plates of calcium carbonate, and with jointed appendages; for example, a crab or an insect.

ASTHENOSPHERE. Upper zone of the earth's mantle extending from the base of the lithosphere to depths of about 250 kilometers beneath continents and ocean basins; relatively weak and mobile layer containing convection cells.

ATOLL. A horseshoe or circular array of islands, capping a coral reef system perched on a volcanic seamount.

AUTOTOMY. Self-crippling by cast-off, or damaged or trapped appendages, such as arms or legs by sea stars or crabs.

BACKING WIND. A wind that shifts in a counter-clockwise direction as follows: southwest, south, southeast, east.

BACKSHORE. Part of a beach that is usually dry and is reached only by storm waves and the highest tides.

BACKWASH. The seaward return of water following the uprush of the waves.

BARRIER ISLAND. An elongate sand island essentially parallel to the shore, commonly having dunes and separated from the mainland by a salt marsh or lagoon.

BASALT. A fine-grained igneous rock, black or greenish black, rich in iron, magnesium, and calcium.

BEACH. A zone of sand or gravel that forms the gently sloping shore of a body of water.

BEACH CUSPS. A series of low mounds separated by crescent-shaped troughs at regular intervals along the foreshore of a beach.

BEACH SCARP. A small, nearly vertical cliff on a beach, caused by wave erosion.

BENTHIC. Organisms that live permanently on or in the bottom of the ocean.

BERM. A low, nearly horizontal portion of the beach or backshore, formed by the deposition of sand or gravel by waves. It marks the ordinary limit of high tides and waves.

BERM CREST. The seaward limit of the berm, which divides the foreshore from the backshore.

BIOTITE. A dark brown to green mineral of the mica group, which forms in igneous and metamorphic rocks.

BIOTURBATION. Stirring up of mud or sand by burrowing organisms.

BIRD-FOOT DELTA. A delta formed at the mouth of a river by outgrowth of pairs of natural levees, formed by the distributaries, making the digitate or bird-foot form.

BLOWOUT. A saucer or cup-shaped hollow formed by wind erosion on a sand dune.

BOULDER. A large, rounded rock that has a diameter greater than 256 millimeters (10 inches).

BYSSUS. A threadlike process used to attach mussels to rocks.

CALCITE. A mineral, composed of calcium carbonate, that is secreted by organisms in shells and forms limestone.

CALDERA. A large, basin-shaped volcanic depression formed by the explosion or collapse of a volcano.

CAPE. A point of land extending into a body of water.

CAPILLARY WAVE. A wave in which the primary restoring force is surface tension on the liquid in which it is traveling.

CENTRIFUGAL FORCE. A force that propels something outward from the center of rotation.

CEPHALOPODS. Benthic or swimming mollusks possessing a large head, large eyes, and a circle of tentacles around the mouth. They include squids and octopi.

CHENIER. An abandoned beach located some distance landward from the shore as the result of the deposition of fine sediment on its seaward side.

CILIA. Hairlike processes of cells that beat rhythmically and cause locomotion of the cells or produce currents in water.

CIRCULAR WAVE. A wave that expands in concentric circles from a point source.

CIRRI. Small, flexible appendages that are present on some invertebrate groups, such as barnacles and annelid worms.

CLAY. A fine-grained mineral formed from the weathering of feldspar in igneous rocks. The grains are smaller than 1/256 millimeter and have a greasy feel.

COAST. Region extending inland from the sea, ordinarily as far as the first topographic change in the land surface. A stretch of the shore together with the land nearby; in general, an area where maritime influences prevail.

COASTLINE. Approximate position of the shoreline, ordinarily as displayed on a map or chart.

COASTAL PLAIN. Low-relief continental plain, adjacent to the ocean and extending landward to the first major change in terrain features.

COASTAL TECTONICS. Movements of the earth's crust that affect the landforms on a coast.

COBBLE. A rock fragment between 64 and 256 millimeters in diameter, larger than a pebble and smaller than a boulder.

COLLAR CELL. Cell in a sponge with a distinct tubular collar about the base of a long, slender, whiplike extension, the flagellum.

COLLISION COAST. An uplifted or block-faulted coast of a continent, facing toward a collision boundary between crustal plates.

COMMENSAL. Having benefit for one but neither positive nor negative effect on another member of a two-species association.

CONDUCTION. The transfer of energy through matter by internal particle or molecular motion without any external motion.

CONGLOMERATE. A sedimentary rock made up of more or less rounded fragments of pebbles, cobbles or boulders cemented by sand or mud.

CONTINENTAL SHELF. The seafloor adjacent to a continent, extending from the low-water line to a change in slope, usually at about 180 meters in depth, where the continental slope begins.

CONTINENTAL SLOPE. A downward slope, averaging about 4 degrees, that extends from the edge of the continental shelf to the continental rise or deep sea floor.

CONVECTION. Vertical circulation caused by density differences within a fluid, resulting in a transport and mixing of the properties of that fluid.

CONVECTION CELL. Convective fluid movement in a mass, with a central portion moving upward and outer regions moving downward. Convection cells in the asthenosphere move lithospheric plates horizontally across the surface of the earth.

CONVERGENT PLATE BOUNDARY. Colliding edges of crustal plates where plate edges may be thickened, folded, or underthrust, or one may be subducted and destroyed.

CORAL REEF. A complex ecological association of bottom-living and attached marine invertebrates, forming fringing reefs, barrier reefs, or atolls.

CORALLITE. Skeleton formed by an individual coral polyp in a colony consisting of walls and septa. The cup of calcium carbonate in which the polyp lives.

CORANGE LINE. A circular or oval line on a tidal chart along which all tidal ranges are equal.

CORE. The central dense zone of the earth, the outer boundary of which is 2900 kilometers below the surface of the earth.

CORIOLIS EFFECT. An apparent force acting on moving particles, resulting from the rotation of the earth. It causes moving particles to be deflected to the right in the Northern Hemisphere and to the left in the Southern Hemisphere.

COTIDAL LINE. A line on a tidal chart that passes through all points where high tides occur at the same time.

CRUST. The outer shell of the solid earth, which ranges in thickness from 5 to 50 kilometers.

CRUSTACEANS. Arthropods that breathe by means of gills or similar structures. The body is commonly covered by a hard shell or crust. The group includes barnacles, crabs, shrimp, and lobsters.

DECLINATION OF THE MOON. The angular distance between the earth's equator and the orbital plane of the moon.

DECOMPOSERS. Bacteria and fungi that aid in the decomposition of the host organism.

DELTA. A wedge-shaped deposit of sand or mud at the mouth of a river or tidal inlet.

DEPOSIT FEEDER. An organism that derives its nutrition by consuming organisms that live in soft sediment.

DEPOSITION. The act of depositing or laying down a layer of mud, sand, or gravel by air, water, or ice.

DESICCATION. Drying out.

DIATOMS. Microscopic single-celled plants that live in fresh and marine water.

DIFFRACTION. Changes that take place in waves when they pass a point of land or the tip of a breakwater. A circular wave is generated at the point that intersects the incoming wave.

DIFFRACTION PATTERN. A wave pattern produced by diffraction at a harbor mouth, characterized by a choppy area, geometric shadow, and phase lines.

DIFFUSION. Transfer of a material (water) or a property (temperature) by eddies or molecular movement. Diffusion causes slow spreading outward from the area of high concentration.

DIKE. A thin sheet of igneous rock, usually basalt, which cuts across the country rock.

DINOFLAGELLATES. Microscopic unicellular animals that have two flagella or whiplike appendages.

DIURNAL TIDES. Daily tides, one high and one low tide each day.

DIVERGENT PLATE BOUNDARIES. Spreading crustal plates where new crustal material is added to a plate. Divergent plate boundaries occur along mid-oceanic ridges.

DOLDRUMS. A narrow belt on both sides of the equator characterized by upward air currents, heat, low pressure, and high humidity.

DRUMLIN. A streamlined hill or ridge of glacial drift with the long axis parallel to the flow direction of a former glacier.

DUNE. A mound, hill, or ridge of wind-blown sand, either bare or covered with vegetation.

EBB CHANNEL. The deepest channel in a tidal inlet, which concentrates the ebb current.

EBB CURRENT. The outward-flowing tidal current in a tidal inlet.

EBB SHIELD. A lip of sand on the landward edge of the flood-tidal delta, which diverts the ebb current.

EBB-TIDAL DELTA. A body of sand deposited on the seaward side of a tidal inlet by the ebb current.

ECHINODERMS. Primitive invertebrates with a fivefold symmetry and a water-vascular system. Group includes sea stars, brittle stars, sea urchins, and sand dollars.

ECLIPTIC PLANE. The plane of the earth's orbit around the sun.

EDGE WAVES. Standing waves with nodal lines perpendicular to the shore. Edge waves are responsible for the rhythmic pattern of rip currents and beach cusps along the shore.

EDGE ZONE CORALS. Reef-forming colonial corals that are fused together by septa that extend outside the corallite.

EPIFAUNA. Bottom-dwelling organisms that live on the top of the sediment.

EQUINOX. The time when the sun crosses the plane of the earth's equator making night and day equal length all over the earth, occurring about March 21 (Vernal equinox) and September 22 (Autumnal equinox).

EROSION. The process by which the earth is worn away by glaciers, rivers, waves, and currents.

ESTUARY. A bay at the mouth of a river where the tide meets the river current.

EULITTORAL ZONE. The zone along the coast between high and low tide.

EUPHOTIC ZONE. The uppermost layer of a body of water, which receives enough light for photosynthesis to take place.

EUSTATIC SEA LEVEL. Pertaining to worldwide and simultaneous change in sea level, such as that caused by melting of glaciers.

EUTROPHICATION. Increase in the nutrient content of a body of water, which depletes the biological oxygen supply.

EXFOLIATION. The breaking or spalling off of concentric sheets of rock from a bare rock surface due to the release of pressure.

FELDSPAR. A group of abundant rock-forming minerals that occur in granite and basalt.

FETCH. The area over which wind blows generating waves during a storm.

FILTER-FEEDER. An organism that feeds by filtering microscopic food particles out of water.

FJORD. A narrow steep-sided inlet of the sea formed by submergence of a mountainous area or excavation by a glacier.

FLAGELLUM. A whiplike process of protoplasm, which provides locomotion for a motile cell.

FLOOD CHANNEL. A channel on the margin of a tidal inlet, which contains most of the flood current.

FLOOD CURRENT. The tidal current that flows into a tidal inlet with the rising tide.

FLOOD RAMP. Seaward-sloping portion of the flood-tidal delta, which is covered by the flood current.

FLOOD-TIDAL DELTA. A body of sand deposited on the landward side of a tidal inlet by the flood current.

FOLIATION. Layered structure in metamorphic rocks, resulting from segregation of light and dark minerals into distinct bands.

FORESHORE. The seaward-dipping zone on a beach between high- and low-tide levels.

FOURIER ANALYSIS. Separation of a curve, such as a tidal curve, into sine and cosine components that can be used to predict the tides.

FREQUENCY. The number of cycles or portions of a cycle within a given time. Frequency is 1/period.

FUNGUS. A group of thallophytes, including mushrooms, mildews, and molds, characterized by absence of chlorophyll and subsisting on dead or living organic matter.

GABBRO. A coarse-grained igneous rock with the same composition as basalt.

GARNET. A purple mineral formed in metamorphic rocks that collects in a band on a beach following a storm.

GASTROPODS. Mollusks that possess a distinct head, generally with eyes and tentacles, and a broad, flat foot; usually enclosed in a spiral shell. Group includes snails, limpets, and slugs.

GEOMETRIC SHADOW. In a wave-diffraction pattern, a line behind a breakwater along which wave height is equal to one-half the incoming wave height.

GEOSTROPHIC WIND. An upper-level wind flowing parallel to the isobars, which is an exact balance between pressure-gradient force and the Coriolis effect.

GNEISS. A banded metamorphic rock with light and dark layers consisting of quartz and feldspar or iron and magnesium minerals.

GRANITE. A coarse-grained igneous rock composed of quartz, feldspar, and biotite or hornblende.

GRAVITY WAVE. A wave whose velocity of propagation is controlled primarily by gravity.

GROUP VELOCITY. The velocity at which a wave group travels. In deep water, it is half the phase velocity. Wave energy travels at the group velocity.

HALOPHYTE. A plant that grows in salty or alkaline soil.

HEXACORALS. Hard corals in which the polyp sits in a cup containing septa in cycles of six. In reef corals, the septa form an edge zone that permits construction of massive encrusting skeletons. Elkhorn, staghorn, and brain corals are included in this group.

HODOGRAPH. A circular or oval plot of tidal current motion in the open ocean.

HORSE LATITUDES. The mid-latitudes where air descends forming subtropical highs. In sailing days, the cargo was lightened by throwing horses overboard in the light winds to decrease the cargo.

HYPHAE. Threadlike elements in bacteria.

IGNEOUS ROCK. Rocks that have formed by crystallization from a molten or partially molten state.

INFAUNA. Animals who live buried in soft sediment.

INFRALITTORAL FRINGE. The lowest level in the intertidal zone, which is exposed only during low spring tide. It is also known as the *Laminaria* zone.

INSHORE ZONE. The zone of the beach profile extending seaward from the foreshore to just beyond the breaker zone.

INTERGLACIAL PERIOD. The warm interval between glaciations, when sea level was higher than present.

ISLAND ARC-SYSTEM. A group of islands having a curving, arclike pattern, convex toward the open ocean. Associated with a trench on the ocean side, indicating a subduction zone.

ISOPODS. Generally flattened marine crustaceans, mostly scavengers, common on sand beaches.

ISOSTATIC REBOUND. The uplift of land that was pushed into the earth by the weight of large glaciers.

KEYSTONE SPECIES. A predator at the top of a food web, capable of consuming organisms of more than one trophic level beneath it.

LAGOON. A bay extending roughly parallel to the coast behind a barrier island.

LAMINARIA. Large algae or kelp that are found in the infralittoral and sublittoral zones.

LANCEOLATE. Shaped like the head of a lance.

LEADING-EDGE COAST. An uplifted or block-faulted coast of a continent, facing toward a subduction zone or collision boundary.

LICHEN. A group of compound plants consisting of fungi in symbiotic union with algae, growing in green, gray, yellow, brown, or black patches.

LIMESTONE. A bedded sedimentary rock, consisting chiefly of calcium carbonate ($CaCO_3$), derived from marine shells or inorganic precipitates.

LITHOTHAMNIA. A red, calcareous algae that forms a ridge on the seaward edge of the reef.

LITHOSPHERE. The outer, solid portion of the earth, which includes the crust.

LITTORAL DRIFT. Sand moved parallel to the shore by wave and current action.

LITTORAL ZONE. The intertidal zone between mean high-water and mean low-water levels.

LONGSHORE CURRENT. A current in the surf zone, moving parallel to the shore and generated by waves breaking at an angle to the beach.

LUNATE MEGARIPPLES. Moon-shaped depressions in a channel, formed by a strong tidal current.

MACROPHAGE. An organism that eats large plants or animals.

MACROPHYTIC ALGAE. Individual algae that are large enough to be easily seen by the unaided eye.

MACROTIDAL. A tidal range greater than 4 meters.

MADREPORE. An external opening that allows water into the water-vascular system of sea stars.

MAGMA. Molten rock in the earth's crust or mantle that crystallizes to form an igneous rock.

MAGNETITE. A black magnetic mineral of iron oxide that collects in bands on storm beaches.

MANTLE. The bulk of the earth between the crust and the core. Also, the tough protective membrane possessed by all mollusks.

MARSH. An area of soft, wet land. Flat land periodically flooded by salt water is called a salt marsh.

MEDUSA. A free-swimming jellyfish having a disc- or bell-shaped body, with dangling tentacles containing stinging cells.

MESOTIDAL. A tidal range between 2 and 4 meters.

METAMORPHIC ROCK. A rock formed in the solid state from a

preexisting rock by the addition of heat, pressure, or charged chemical fluids.

MICROPHAGE. An organism that eats microscopic plants or animals.

MICROSEISMOGRAPH. A seismograph that is used for detecting small earth tremors. Used for measuring waves as they cross the surf zone.

MICROTIDAL. A tidal range of less than 2 meters.

MIDLITTORAL ZONE. The middle intertidal zone that extends from the upper limit of laminarians to the upper limit of the barnacles.

MIXED TIDES. A combination of diurnal and semidiurnal tides that gives a higher-high, a lower-high, a higher-low, and a lower-lov tide each day.

MONSOON. Seasonal winds that blow for six months from one direction, then reverse and blow for six months from the other direction.

MORAINE. An accumulation of poorly sorted sediment, carried by a glacier, which is deposited in a ridge along the end and sides of a glacier.

NEAP TIDE. The lowest tidal range during the lunar cycle, when the sun and moon are at right angles to the earth. It occurs twice each month, at the first and third quarters of the moon.

NEMATOCYSTS. Stinging cells contained in the tentacles of jellyfish and corals.

NODE. That part of a standing wave where the vertical motion is least and the horizontal motion is greatest. It occurs at intervals of half a wave length starting ¼ wave length from a reflecting wall.

NUDIBRANCHS. Gastropods in which the shell is entirely absent in the adult. The body bears projections that vary in color and complexity. Commonly called sea slugs.

OCEANIC TRENCH. A deep linear trench formed along a subduction zone and adjacent to an island arc along a collision plate boundary.

OCTOCORALS. Soft corals or gorgonians in which the tiny polyps are arranged on a flexible skeleton. Sea fans and sea whips are included in this group.

OFFSHORE ZONE. The comparatively flat portion of a coastal profile extending seaward from the breaker zone to the edge of the continental shelf.

OLIVINE. A dark green rock-forming mineral found in basalt and gabbro.

OOLITE. A spherical sand-sized grain of calcium carbonate with concentric shells, formed by precipitation in the wave-agitated zone along tropical reefs.

OPERCULUM. A lid or covering used by a snail to seal its shell against intruders or to prevent loss of moisture.

OUTWASH PLAIN. A plain composed of sand and gravel deposited by meltwater streams beyond the margin of an active glacier.

PEDICELLARIA. Tiny jawlike appendages on the outer surface of echinoderms, used for walking, digging, or self-defense.

PELECYPODS. Mollusks with two valves or shells, a powerful foot for burrowing or swimming, and siphons for eating and breathing. Group includes mussels and clams.

PERIGEE. The point in the orbit of the moon that is nearest to the earth.

PERIHELION. The point in the orbit of the earth when it is nearest to the sun.

PERIOD. The length of time it takes for one wave length to pass by a stationary point.

PHASE LINE. A line on a diffraction pattern along which circular and incoming waves cross. Wave height along phase lines is about 1.2 times incoming wave height.

PHASE VELOCITY. The speed at which an individual wave travels on the ocean surface.

PHYLLITE. A metamorphic rock derived from shale, which is intermediate in grade between slate and schist. Mica minerals impart a silky sheen to the surface.

PHYTOPLANKTON. Plants that drift with the ocean currents.

PLANKTON. Plants or animals that drift with the ocean currents.

PLATE TECTONICS. The theory that accounts for earthquakes, volcanoes, mountain building, and other manifestations of plate movement, with sea floor spreading and horizontal movement of lithospheric plates on a fluid asthenosphere.

PLATFORM REEFS. Circular or oval-shaped reefs that develop on the continental shelf and are protected from direct wave action by wall reefs.

PLUNGING BREAKER. A breaking wave that tends to curl over, forming a tube as it breaks. Characteristic of a medium nearshore slope.

PLUTONIC ROCK. An igneous rock that crystallized at great depth within the earth and therefore has large crystals formed by slow cooling. Granite and gabbro are plutonic rocks.

PODIA. Small tubes closed at the tips and extending singly or in groups from sides of the radial canal in echinoderms.

POLAR FRONT. The boundary between cold, dry polar air masses and warm, moist maritime air masses, along which storms travel from west to east.

POLDER. A land area that has been reclaimed from the ocean; often separated from the ocean by dikes, then drained to form farms or cities.

POLYP. An individual of a solitary coelenterate or one member of a coelenterate colony. Corals and sea anemones are examples of polyps.

PRECESSION OF THE EQUINOX. The earlier occurrence of the equinoxes in each successive sidereal year because of slow retrograde motion of equinoctial points along the ecliptic, caused by the combined action of the sun and moon. A complete revolution of the equinoxes requires about 26,000 years.

PREDATOR. An organism that consumes another organism. (Carnivores and herbivores are predators by this definition.)

PRESSURE-GRADIENT FORCE. A force acting on a body, which moves from an area of high pressure to an area of low pressure. In the atmosphere, the pressure-gradient force is at right angles to the isobars, from high to low pressure.

RADIAL CANAL. In echinoderms, a tube extending from the ring canal to the podia, which is filled with water.

RADULA. Rasplike organ used by many gastropods in feeding.

REFLECTED WAVE. A wave that is returned seaward when it is reflected by a steep beach, seawall, or other reflecting surface.

REFRACTION OF A WAVE. The process by which the direction of a wave is changed as it moves into shallow water at an angle to the bottom contours. The wave crest bends toward alignment with the underwater contours.

RHIZOMES. Horizontal rootlike stems that produce roots below and send up shoots above. Beach grass spreads by sending out rhizomes.

RIDGE. An elongate rise in elevation. In the middle of the ocean, a ridge is formed at the spreading boundary of plates. Along a beach, a ridge is a shallow bar that is separated from the shore by a runnel.

RIFT VALLEY. A trough formed by faulting along a spreading boundary between plates.

RILL MARK. A small groove, furrow, or channel in mud or sand on a beach, made by the backwash of waves, dropping tide, or draining ground water from the foreshore.

RING CANAL. A circular tube connected with radial canals in the water-vascular system of echinoderms.

RIP CHANNEL. A channel or trough between adjacent longshore bars, following the path of a rip current.

RIP CURRENT. A strong, narrow current that crosses the surf zone. A rip current is fed by feeder currents that run parallel to the shore and turn seaward. The current ends in a rip head outside the surf zone.

RIPPLE MARK. A small sand ridge on a bottom, formed by waves or currents. Wave ripples have a symmetrical profile and current ripples are asymmetrical with the steep face heading downcurrent.

RUNNEL. A shallow trough roughly parallel to the beach, bounded on the seaward side by a ridge. Ebb tide flows out the runnel between ridges, forming a rip channel.

SALINITY. A measure of the quantity of dissolved salts in seawater. It is defined as the total amount of dissolved solids in seawater in parts per thousand (0/00) by weight when all carbonate is converted to oxide, the bromide and iodide are converted to chloride, and organic matter is oxidized.

SAND. Loose grains of sediment, usually quartz, feldspar, or calcite, ranging between 1/16 and 2 millimeters in diameter.

SAND WAVES. Large undulations on a sand bottom, which are formed by strong currents.

SCAVENGERS. Organisms that feed on dead organic matter.

SCHIST. A medium- or coarse-grained metamorphic rock with secondary mica grains, aligned parallel to foliation planes or schistosity. Formed by subjecting shale to heat and pressure.

SCYPHOZOANS. Jellyfish in which the medusa stage is well developed and the polyp stage is minimized.

SEA BREEZE. A breeze that blows from the sea toward the land during the afternoon, when air heated over the land expands and rises, forming a partial vacuum.

SEA CUCUMBERS. Echinoderms of the class *Holothuroidea,* which lie on their side with the mouth at one end and the anus at the other.

SEA STARS. Star-shaped echinoderms with arms in multiples of five arranged around a mouth on the bottom side. Commonly called star fish.

SEA URCHINS. Spiny echinoderms that resemble sea stars, with their arms gathered at the top and covered with spines.

SEDIMENT. Loose, unconsolidated material deposited by waves, rivers, glaciers, or wind.

SEDIMENTARY ROCK. A rock formed by the accumulation of pebbles, sand, silt, mud, or shells, which were deposited by waves, currents, rivers, glaciers, or wind. Sedimentary rocks include sandstone, limestone, shale, and conglomerate.

SEICHE. A standing wave in an enclosed body of water, such as a lake or bay, caused by a sudden rise in barometric pressure during the passage of a cold front.

SEMIDIURNAL TIDE. A twice-daily tide that has a period of about 12 hours and 25 minutes between successive high or low tides.

SEPTA. Vertical plates or partitions arranged radially between the axis and wall of a coral. Septa alternate in position with mesentaries or folds in the coral polyp.

SHELAE. Large claws on crabs and other Crustacea.

SHOALING WAVES. Waves that enter shallow water with a decrease in wave length and phase velocity, and a decrease followed by an increase in height, while period remains constant.

SHORELINE. The line of demarcation between the water and the exposed beach.

SILT. Sedimentary particles between sand and clay, which range in size between 1/16 and 1/256 millimeter.

SLIDING-EDGE COAST. A boundary between adjacent lithospheric plates, formed by one plate sliding past another.

SOLSTICE. The two times of the year when the sun is at its greatest distance from the celestial equator. The summer solstice in the Northern Hemisphere is on or about June 21, and the winter solstice is on or about December 22.

SPICULES. Small needlelike rods, or fused clusters of rods, that form the skeletal framework of some sponges.

SPILLING BREAKER. A wave that starts to break along its crest and continues to break as it moves toward the shore. Characteristic of a low nearshore slope.

SPILLOVER LOBE. A small lobe of sand deposited on the edge of a flood-tidal delta by an ebb current.

SPIT. A small point of land projecting into water or an exposed sand bar that is attached to the land at one end. Sand spits are formed by longshore currents moving past a point of land.

SPREADING BOUNDARY. A boundary between adjacent lithospheric plates where the plates are moving apart and new crustal material is plastered to the edges of the plates.

SPRING TIDE. The largest tidal range, occurring twice each month about three days after the new and full moons, when the earth, moon, and sun are in a line. The highest spring tide occurs at the new moon when the sun and moon are on the same side of the earth.

SPONGIN. Soft, flexible organic material that forms the body of many sponges.

SPUR AND GROOVE. The upper part on the windward edge of a reef, composed of buttress-shaped coral ridges with deep, narrow channels between them.

SQUALL. A small, intense storm with high winds and heavy rain of short duration. A line squall usually forms about 200 kilometers ahead of a strong, cold front and moves forward at 20 to 40 kilometers per hour.

STANDING WAVE. A wave in which the surface of the water oscillates up and down between fixed points called nodes. The points of maximum rise and fall are called antinodes. Standing waves are caused by the interaction of incoming and reflected waves.

STIPE. A stalk or slender support of a plant.

STORM SURGE. A rise in water level along a coast, caused by strong onshore winds or hurricane.

STRATOSPHERE. The upper atmosphere above the troposphere, which is not affected by daily changes in the weather.

SUBDUCTION ZONE. An inclined plane descending away from a trench, separating a sinking oceanic plate from an overriding plate. Formed by colliding plates and associated with earthquakes and volcanoes.

SUBLITTORAL. A zone extending from low tide to the edge of the continental shelf.

SUBSTRATE. A surface upon which organisms are attached or move about.

SUPRALITTORAL FRINGE. An ecologic zone, above the midlittoral zone, from the highest occurrence of barnacles to the top of the splash zone. Often called the black zone or periwinkle zone.

SURGING BREAKER. A wave that surges up the beach without spilling or breaking and is reflected out to sea on a steep foreshore.

SUSPENSION FEEDERS. Organisms that feed by capturing particles suspended in the water column.

SWASH MARKS. Arcuate lines of sand or shell debris left by the uprush of a wave.

SWASH ZONE. The narrow zone on the foreshore where water rushes up and back on the beach.

SWELLS. Long period waves with a smooth profile that move out of the storm area and travel great distances across the ocean.

SYMBIOTIC. Two species of organisms that live together to their mutual advantage.

TECTONICS. Rock structures and landforms resulting from movements of the earth's crust. Forces producing earthquakes, volcanoes, and landslides.

TERMINAL LOBE. A crescent-shaped sand bar that extends across the mouth of a tidal inlet at the seaward margin of the ebb tidal delta.

TERRESTRIAL. Related to land, such as terrestrial organisms versus marine organisms.

TIDAL AMPLITUDE. Half of the tidal range, or the vertical distance from mean sea level to high or low tide.

TIDAL COMPONENTS. Individual sine or cosine curves assigned to movements of the sun or moon around the earth. The tidal components are added together to predict the tides at a location.

TIDAL CURRENTS. Ebb and flood currents resulting from the passage of the tidal wave. Flood currents occur with a rising tide, and ebb currents with a falling tide.

TIDAL PHASE. The relative initial time for each tidal component at a coastal location.

TIDAL PRISM. The volume of water that moves into or out of a tidal inlet during each tidal cycle.

TIDAL RANGE. The vertical distance between high and low tide in a tidal cycle.

TRADE WINDS. Steady winds north or south of the equator that blow from the subtropical high toward the equatorial low. The trade winds blow from the northeast in the Northern Hemisphere and from the southeast in the Southern Hemisphere.

TRAILING-EDGE COAST. A gently sloping coast with barrier islands and a broad continental shelf, on the trailing edge of a continent, facing toward a spreading zone along a mid-oceanic rise.

TRANSFORM FAULT. A fault connecting the offset portions of a mid-oceanic ridge along which crustal plates slide past each other.

TRILOBITES. Extinct arthropods with three lobes—a center or axial lobe, and two side or pleural lobes. They had well-developed eyes and reigned supreme for almost 300 million years in the Paleozoic oceans.

TROUGH. A linear depression generally parallel to the shore, excavated by wave action and longshore currents in the breaker zone. The longshore bar rises a meter or more seaward from the trough. Also, the lowest point on a wave profile between wave crests.

TSUNAMI. A long-period, high-velocity sea wave produced by submarine earthquake, volcanic eruption, or landslide.

TUBE FEET. Small tubes or podia that are closed at the tip and extend from the radial canal in echinoderms, including sea stars and sea urchins. The tube acts as a suction cup, allowing the sea star to walk.

TUBERCLE. A small rounded projection on the surface of an animal or plant.

UPWELLING. The process by which water rises from a lower to a higher depth, usually the result of a divergence of surface currents or a strong current moving parallel to the shore that is diverted offshore by the Coriolis effect.

VEERING WIND. A wind that shifts in a clockwise direction as follows: northwest, north, northeast, east.

VISCERAL MASS. The soft interior organs in the cavities of the body, including the heart, lungs, stomach, and intestines.

WALL REEFS. Elongate coral reefs along the wave-exposed portion of the Great Barrier Reef in Australia. Usually lined with spurs and grooves.

WASHOVER FAN. A fan-shaped deposit of sand laid down on the backside of a dune by wave surges that pass through low breaks in the dune line.

WATER-VASCULAR SYSTEM. The water circulation system in echinoderms, which includes the ring canal, radial canals, and podia or tube feet.

WAVE CREST. The highest point along a wave profile.

WAVE HEIGHT. The vertical distance from the bottom of the trough to the top of the crest on a wave.

WAVE LENGTH. The horizontal distance from one point along a wave profile to an equivalent point on the following wave, usually measured from one wave crest to the next.

WAVE ORBIT. The circular or oval path followed by water particles in a wave.

WAVE RAY. A line perpendicular to a wave crest, indicating the direction a wave advances. Wave rays converge on a headland and diverge in a bay.

WAVE SPECTRUM. A graph showing the distribution of wave energy (wave height squared) for different wave periods. A characteristic wave spectrum can be plotted for wind blowing with a given wind speed, fetch, and duration.

WAVE STEEPNESS. The ratio of wave height to wave length.

WESTERLIES. Mid-latitude winds between the subtropical high and the polar front, which blow steadily from the west.

ZOOPLANKTON. Animals that drift with the ocean currents.

ZOOXANTHELLAE. A group of yellow or brown dinoflagellates living symbiotically with one of a variety of invertebrates including corals and clams.

Index